Edward C. Mann

A Treatise on the Medical Jurisprudence of Insanity

Edward C. Mann

A Treatise on the Medical Jurisprudence of Insanity

ISBN/EAN: 9783337314422

Printed in Europe, USA, Canada, Australia, Japan

Cover: Foto ©berggeist007 / pixelio.de

More available books at **www.hansebooks.com**

A

TREATISE

.

ON THE

MEDICAL JURISPRUDENCE

OF

INSANITY.

BY

EDWARD C. MANN, M. D.,

Author of "A Manual of Psychological Medicine."

MATTHEW BENDER.
LAW BOOK PUBLISHER.
1893.

.

BOYD'S ALBANY PRINTING CO.. PRINTERS.
ALBANY, N. Y.

Dedication.

TABLE OF CONTENTS.

I desire to give credit to my work, " A Manual of Psychological Medicine," for remarks on the diagnosis of insanity, unsoundness of mind, states of unconsciousness, etc.   E. C. M.

INTRODUCTION.

THE PHYSICAL BASIS OF VOLUNTARY ACTION, MEMORY, EMOTION AND THOUGHT.

A study of the physiology of the nervous system shows us that nerve-force is generated in or by the vesicular neurine, and that the tubular or fibrous neurine conducts it. But, what is the nerve-force of the brain? And in what manner is its vesicular neurine active? The result of its activity we know, namely, the ability to receive sensations of all kinds; the power of comparing these sensations and of storing them for comparison; the power of combining these sensations in new arrangements, of imagining — not, indeed, new sensations — but combinations of them; the power of feeling emotions and propensities.

The activity of the vesicular neurine of the brain is the occasion of all these capabilities and powers. The little cells are the agents of all that is called mind, of all our sensations, thoughts and desires; and the growth and renovation of these cells is the most ultimate condition of mind with which we are acquainted. If there are more profound conditions they are beyond our knowledge. Now, how is the nutrition of the brain-cell effected?

The grey substance of the brain contains millions of vesicles, lying in a semi-fluid granulated substance (stroma), and bound together by a minute network of capillary blood-vessels and fine areolar tissue. Now, the fundamental truth of physiology being the activity of the cell, and this activity being accompanied by its decay, and demanding its renovation : the important points in

the relative position of the brain-cell are, first, its relation to the nerve fibre, from which it receives and to
which it conveys impressions, the taking and giving of
which are the main causes of its exhaustion ; second, its
relation to the blood capillary, which exudes a plasma
in which the cell is bathed and renovated, and from
which new cells are formed to replace those which are
finally exhausted. Respecting the first of these relations, so far as the individual cell is concerned, it would
appear, that morbid results could only arise from stimulation so excessive as to hasten the progress of decay
beyond the power of reparation. Respecting the secondary relation, numberless circumstances may occur to
interrupt or prevent the growth or reparation of the cell.
All states, either physiological or pathological, of the
brain-cell, are occasioned by influences impressed upon it
either by the nerves or blood vessels, with which it
stands in such intimate relation. Whether any changes
can be self originated is doubtful. The laws of its life,
transmitted to the cell from the parent organism, include,
indeed, the conditions of perpetual change, but the cause
of change must be sought for in the nerve, or capillary.

  Voluntary action.—What happens in the simplest
form of voluntary action ? A baby sees what it wants
and seizes it. There is a particular impression on the
sense of sight, a conduction of the molecular motion
caused thereby to a special nerve-centre, and the consequent excitation of a special perception as the ingoing
process ; then, as the outgoing process, the transmission
of liberated energy along motor nerve to muscle and a
consequent adaptive act — what we call a reflex action
in the mental plane. Voluntary action is a power of
better quality and higher dignity in man than in the
animal or the baby, but even in its first stage of evolution the animal and the baby soon learn to voluntarily
restrain themselves from doing what their first impulse
was to do, of two courses to choose the best, the path of
prudence or duty in preference to the path of natural

proclivity. They are taught by adults, and where the animal or the baby withstands the impulse to do what it wants to do, there is, along with the special perception, the immediate stimulation of the associated nerve-centre that has suffered and registered the memory of the suffering ; and the consequence is, the resistance to, or inhibition or check of the instant impulse and the prevention of the movement. It is the physical basis of morals, I think, that one of two associated nerve-centres may be excited to check or inhibit the other. It is by virtue of this physiological property that structure is moulded along the lines of function, and that the ease of performance which we call habit is acquired. A voluntary action, then, is a motion or group of motions accompanied or preceded by volition, and directed toward some object. Every such action comprises the following elements : knowledge, motion, choice, volition, intention and thoughts, feelings and motions adapted to execute the intention. These elements occur in the order in which I have enumerated them. Suppose one of you about to act. Your knowledge of the world in which you live, and of your own power, assures you that you can, if you like, do any one or more of a certain number of things, each of which will effect you in a certain definite way, desirable or undesirable. You can speak or be silent. You can sit or stand. You can read or write. You can keep quiet or change your position to a greater of less extent and by a variety of different means. The reasons for and against these various courses are the motives. They are taken into consideration and compared together in the act of choice, which means no more than the comparison of motives. Choice leads to determination to take some particular course, and this determination issues in a volition, a kind of crisis, of which every one is conscious, but which it is impossible to describe otherwise than by naming it, and as to the precise nature and origin of which many views have been entertained which I need not here discuss.

c

The direction of conduct towards the object chosen is
called the intention or aim. Finally there takes place a
series of bodily motions and trains of thought and feeling fitted to the execution of the intention. I think this
is a substantially correct account of normal voluntary
action. It would be difficult to attach any meaning to
the expression "voluntary action," if either motive or
choice, or a volition, or an intention, or actions directed
towards its execution, were absent, though they may not
always be equally well marked.

Mental disease, or disease of the brain, powerfully
affects, or may affect, the knowledge by which our
actions are guided, the feelings by which our actions are
prompted, the will by which our actions are performed,
whether the word will is taken to mean volition or a
settled judgment of the reason acting as a standing control on such actions as relate to it. The means by which
these effects are produced are unnatural feelings ; delusions or false opinions as to facts ; hallucinations or
deceptions of the senses ; impulses to particular acts or
classes of acts ; and in some cases, it is asserted, a specific
physical inability to recognize the difference between
moral good and evil as a motive for doing good and
avoiding evil.

The brain and nervous systems are the organs by which
all mental operations are conducted. When a man
either feels, knows, believes, remembers, is conscious of
motives, deliberates, wills, or carries out his determination, his brain and his nerves do something definite,
though what that something is, what parts of the brain
are specially connected with particular mental functions,
by which part a man remembers, by which he conceives,
and how any part of the brain acts when any of these
operations is performed, no one knows. All that we
know is, that one set of nerves carry to the brain a
variety of impressions of external objects and occurrences, that these impressions excite emotions which
affect many parts of the body in different ways, and

which in particular affect the brain ; that the brain in some manner deals with the impressions, whether of perception or of emotion, which it receives, during which process the man is conscious of what we describe as emotion, motive, deliberation and choice, and at the moment when the man is conscious of volition, some discharge from the brain, through a different set of nerves from those which convey impressions to it, acts on the various parts of the body in such a manner as to cause those groups of bodily motions which we call voluntary actions.

The functions of feeling, knowing, emotion and willing are mental functions, and, as we have seen, the ultimate condition of mind with which we are now acquainted consists in the due nutrition, growth and renovation of the brain cells. We have seen that the brain cells are nourished from the blood, and this nutritive renovation takes place probably, principally, if not entirely, during sleep, and anything that interferes with the uniform and healthy interchange of nutritive plasma passing from the vessels to the cells, and of the fluid cell contents in a state of involution or degenerative metamorphosis passing from the cells to the vessels, deranges the intimate connection between the nervous and vascular systems through which their most important functions are performed, producing at once grave disturbances of the nervous system, which may affect voluntary action and also memory, emotion and thought. In fact, the elementary disturbances of the brain functions which we meet with in disease of the mind, involve processes in the emotional sphere : processes in the sphere of the conceptions, comprising the reason and memory, and processes in the psycho-motor sphere, the impulses and the will. The result of these morbid processes is to induce morbid delusive conception or perception of subjective origin, causing change of mental character as compared with former self or normal ancestral type, through organic conditions originating in disease within the system, external motives playing but a secondary part when they influence at all the mental conduct. To have normal voluntary action, memory, emotion and thought, we must have a normal cerebrum, with a proper conformation and development of hemispheres ; normal coverings or membranes of the brain, with blood vessels of the normal size ; the convolutions of the brain must be developed sufficiently; we must have, in short, freedom from any and all abnormalities, either of structure or function, of the centric nervous system. Any disease, malformation or interference with these parts by mechanical, chemical or vital causes, may produce disease and interfere with voluntary action, memory, emotion and thought. Criminal impulses may easily haunt a mind naturally innocent and pure, when the mental functions of feeling and knowing, emotion and willing, are not performed in their regular and usual manner by reason of diseases of the nervous system. It is a very simple thing to get up a false action of conception or judgment, a defective power of the will or an uncontrollable violence of the emotions and instincts ; first, by primary disease or defective development of the encephalic centres ; second, by disorder of, or developmental changes occuring in other organs than the encephalic centres, and third, by stimulants and narcotics.

Among the predisposing causes which disturb memory, emotion and thought, we find hereditary predisposition, the great mental activity and strain upon the nervous system that appertain to the present age and state of civilization, great difference of age between parents, influence of sex, of surroundings, pre-natal causes, physiological crises, epilepsy and intemperance. Among the exciting causes may be mentioned trouble and excessive grief, intemperance, excessive excitement of whatever kind, epilepsy, disordered physical functions, fevers, injuries to the head and spine, sunstroke and overwork. I think fifty per cent. of all the disturbances of memory, emotion and thought may be traced directly or indirectly to intemperance.

To have voluntary action, memory, emotion and thought of the highest type, races should strive to get a physique and a stability of nerve tissue capable of meeting successfully the demands that climate and civilization make. Great attention should be paid to environment, diet, fresh air, exercise, proper rest, sleep and tranquility of life. Respecting emotion and thought, there is a realm of inquiry of unprecedented interest to the student of psychology and of immense importance in many practical relations of life. I refer to the effect of the hereditary taint of insanity on man and his offspring. The brain comes into the world with the same imperfections and deficiencies, the same irresistible tendencies to disease or perversity of action, which have long been observed in regard to other organs. We must necessarily take this into account in forming theories of human action, and it will throw new light on many a dark problem of human conduct.

Emotion and thought, conduct and responsibility, are seriously affected, and responsibility in particular may be impaired, by that cerebral condition—neither of health nor of disease, as these terms are usually understood—which is produced by tendencies to disease or ancestral vices. There is a prevalent mistake of regarding no disease as hereditary that does not descend fully formed directly from parent to child. This is not so, for the force of inertia often causes disease, and especially mental disease, to remain latent for an entire generation and probably more. To lie concealed, however, does not mean necessarily to be destroyed. Disease is an ultimate product, elaborated from simpler elements during a period that may embrace more than one generation ; its essential element, considered in any stage of its progress, is imperfection, defect, abnormal depreciation, to be manifested under the process of hereditary transmission in every variety of form. This leads us to a correct theory of the production of disease of the mind. Under the adverse influences of a highly civilized condition, the

cerebral system suffers in common with the others; while
the signs of such suffering will generally be found only
in the mental manifestations, varying all the way from
some slight peculiarity or anomaly of character to the
gravest moral or intellectual impropriety. The defect,
under the predominant influence of a different blood,
may not be witnessed in any succeeding generation.
More frequently, however, even in spite of this conservative influence when present, it does make its appearance
again, in one or more of the descendants, in forms more
or less severe than the original.

Memory, emotion and thought and also character and
conduct are determined, in a great degree, by the original
constitution of the brain and nervous system, and I feel
sure that, as you all are led by the teachings of science
and the stern facts of observation, you will be disposed
to make a proper account of those cerebral qualities
which imply a deviation, of some kind or other, from
the line of healthy action. That it will be in accordance
with your philosophy to see in them an explanation of
those strange and curious traits which are utterly inexplicable on the principles that govern the conduct of
ordinary men, and I feel sure that you have been led far
enough in the paths of science to find in them a clew to
some of the mysteries of human delinquency. Only the
more demonstrative forms of mental disease are supposed
to be capable of affecting the legal responsibility of men,
but you all well know that a person who, in most respects,
is rational and observant of the ordinary proprieties of
life, may be so completely under the influence of disease
as to be irresponsible for many of his acts. The course
of emotion and thought, the sense of moral distinctions,
the actual conduct, may all be greatly affected by the
influence of such imperfection. We are bound by a
sense of justice and the claims of science to make some
account of it in forming our estimates of character and
fixing the limits of responsibility; and knowing that an
individual is descended from a line of progenitors

abounding in every form of nervous disorders, shall we think it strange that some portion of it has come to him, and knowing that the quality of the brain is necessarily affected by such disorder, shall we not seek in this fact for an explanation of what we could not explain upon any ordinary principles of human conduct? The hereditary taint of insanity appears in various stages of progress, from the lowest to the highest grade of intensity, and under a variety of forms and aspects. Their effect on the mental capacity and vigor and on the complexion of the moral sentiments and determinations; their connection with the habitual and transitory impulses, with the power of resisting evil and pursuing good, are points we need to study more thoroughly in forming our estimates of character, especially in reference to moral and legal responsibility. They are agencies which should no less engage our attention and be no less effective than education, social influences or mental endowments. Defect of brain, necessarily in one way or another, affects its functions, causing changing and misleading subjective impressions of the person affected, coupled with the resultant change of conduct or of reasoning, or both. There is morbid delusive conception or perception of subjective origin, causing change of mental character as compared with former self or normal ancestral type, through organic conditions originating in disease within the system, external motives playing but a secondary part when they influence at all the mental conducts. Change of character is the ultimate symptomatic expression of mental disease, change of mental conduct the immediate, and repetitions of conduct make character.

We have made an attempt at an explanation of the physical basis of voluntary action, memory, emotion and thought, and finally have tried to explain some of the great underlying truths that govern mental pathology. I have been guided by both an active and a thoughtful acquaintance of the subject, the result of experience in the field of diseases of the nervous system, and I crave indulgence for any shortcomings.

REGARDING MEDICAL EXPERT WITNESSES.

Regarding medical expert witnesses, we would recommend that no person should be permitted, by the trial judge, to testify as an expert who has not been particularly conversant with that department of knowledge to which such facts as he is to testify concerning, belong. Any physician, for instance, may testify about the circumstances of a wound, but only a practical operating surgeon can properly determine whether a given wound will be necessarily fatal. Conduct and conversation may be correctly related by any good observer, but only an alienist physician, of practical experience with the insane, can say with authority whether the mental manifestations described, are indications of insanity. We shall indeed never see absolute, unexceptional agreement, because experience demonstrates that all men do not see and hear the same things exactly alike. Therefore, we cannot have unanimity in matters of opinion any more than in matters of fact. An expert is in court to tell the significance of certain facts that have appeared in evidence. *Ergo.* No man should be permitted to testify as an expert who has not made that class of facts his particular study, and who has not had an opportunity of seeing them displayed on a large scale. All have not had precisely the same experience ; therefore, you will never see invariable unanimity on the witness stand, any more than you will see one judicial decision express the united opinions of the full bench. I distrust opinions generally when they are unanimous, as I expect to find that several men have unthinkingly followed the first of their number who has expressed an opinion. Discrepancy often implies intelligent thought, instead of incompetence. One great evil is that there are some men who, because they have a little actual knowledge on some specialty, claim credit for a great deal which they do not possess, and rush to the witness stand to assume a duty for which they are entirely incompetent, and, in their character of experts, their opinions may have the same weight with the jury as those of better men. The question should be for the lawyer preparing the case, who desires the services of an expert: "Has the man I intend to call, as expert in a given department, given his time and attention entirely to such pursuit?" If not, reject him, and find one who has done so, if you expect skilled testimony that will voice science.

A physician who studies his case, who is conscientious, and who commits himself to no doubtful theories, need have no fear of cross-examination. An expert who goes upon the witness stand and testifies to the truth, and nothing but the truth, so far as his opinion is concerned, would very much prefer to be cross-examined by a lawyer who has studied the subject than by one who has not studied it. As a rule, expert witnesses are treated with respect and consideration while under cross-examination, and it is rarely that lawyers abuse cross-examinations. The present system of calling expert witnesses may have some evils, but, everything considered, it is problematical if any other system that can be suggested would not have equally great disadvantages attending it.

It would be a measure of doubtful propriety to inaugurate any system that would obviate the necessity or do away with the right of cross-examination in open court in the presence of the jury. Every person accused of crime has a constitutional right to be confronted by his witnesses in open court, and has a right to have their truth sifted by cross-examination.

PREFACE.

I HAVE aimed to present the subject of the Medical Jurisprudence of Insanity in a manner to bring the science up to its standard. Public policy may require a rule different from that which would meet the approval of the most advanced scientific minds. Rules of law must not only be framed with a view to protect the accused, but also to protect the community, and therefore I have endeavored to be as conservative as possible, while adhering strictly to scientific truth. I claim that a scientific truth can never be a dangerous doctrine. Our present theories of crime and criminals will soon have to be abandoned. The advances of scientific knowledge in the field of physiology and psychology, sociology and anthropology contradict and give no support to the present views. A new land of science is waiting to be explored. Criminals are the result of natural causes and reversionary types. Pre-natal environment plays an important part in the production of criminals, the insane, the idiot and the epileptic. In studying heredity we find that there is an innate predisposition and an element of contagion from the surroundings. Motor activity is prominent, and physical insensibility being below the normal, in most cases punishment by pain has less influence on the criminal. Moral insensibility and paralysis of all the higher brain forces are found, suggesting an arrested development. We must study the forces and laws which govern crime and the criminal, and adopt measures of treatment that will antagonize the cause of crime. We find distinct cranial and cerebral characteristics in the criminal. Although the average size of crania differs but little from normal people, there is a lack

of symmetry, and a defective development. In the face
we see a large, prominent, lower maxillary bone, or a thin,
retreating jaw. We also find high cheek bones, large,
coarse ears, pallid skin, precocious wrinkles, anomalies
of the hair, and a peculiar physiognomy in which the
eye is most prominent. In large numbers of criminals
there is very striking resemblance. Hearing and the
sense of smell are below the average, while the sense of
sight is very acute. Physiological psychology opens its
researches into the beginnings of life and activity.
Arrested development of brain is accompanied by arrested
development of mind, and we need to open institutions
for the study and treatment of the children of criminals,
so that we may combat degeneration of the tissues of the
brain, and thus prevent the child, so far as possible, from
growing up a criminal, and by the influence of a good
environment train and educate such children mentally,
morally and physically. Degeneration of the tissues of
the brain must always be associated with impaired function. Such study of such children demands observation
and careful, trustworthy records. We must collect data
in child growth of the children of criminals, if we are
going to solve the problem satisfactorily. The senses,
motor activity, the will, the sentiments, consciousness of
the ego, memory, reasoning, the musical sense, language
and expression must all be studied, and records kept
carefully. The child's brain grows as much in the first
year as in the whole of after life ; therefore, if the child
can be taken at birth out of a criminal environment and
placed in a different one, the possibilities would be great
for rescuing that child from the career of a criminal and
to grow up to marry and breed its like, and for bringing
up these children with comparatively normal functions
of feeling and knowing, emotion and willing. The second
generation would be better yet, and in the third we think
all traces of a criminal nature would be obliterated. Is
this not worth the attention of the Commonwealth ?
The subject of Education and Environment *vs.* Heredity,

Criminality, etc., does not receive its due attention. A low grade of brain development is susceptible of much modification by judicious training. Brain structure bears a very close relation to mental states and conduct, and the State should prevent the marriage of criminals. Intemperance in parents may produce a defective brain in the offspring, and owing to the correlation of morbific forces, the child may either become criminal, insane, idiot or epileptic, as the case may be ; therefore, too much attention cannot be given to teaching the physiological action of alcohol on man and his offspring. The habitual criminal is what he is, through a perverted and defective organism. He is permeated by vicious heredity, and, in addition to all this, all his education is, during the most actively formative years, in the direct line of vice. In addition to this, society itself is firmly opposed to the reform of the criminal, and this opposition takes the shape of an ostracism which makes his every sentence perpetual and which excludes him from the hope of public clemency.

PRE-NATAL INFLUENCES IN THEIR RELATION TO CRIME AND CRIMINALS.

Intensely active molecules of the mother, imperceptible to sense, veritably extra-sensual, are the foundation of all the unborn babe's visible matter. Is it then at all inconceivable that in these physico-mental functions, which of all the operations in nature known to us are the finest and most subtle, there are agencies so fine, so little material, as to be unappreciable in themselves and known only by their effects, viz., the future life of the unborn child ? There is no difficulty in a mother's enunciating lofty. general, moral principles to her son, but the son who will be most apt to apply the principles to the particular case, which is the great difficulty in morals, will be the one whose mother, while carrying him while yet unborn, gave him a code of exact rules to help him at such practical junctures in his future life,

by asking God, her Father, each day of her pregnancy, to guide and direct her in all she thought and
said and did, and the repetitions of her conduct will not
only mould and make her own character, but not less
surely that of her unborn infant. A great deal is said
about environment. Environment can do a great deal
for a boy or a girl, a man or a women; but always must
there be something akin within to vibrate in sympathy
with the quality of the power without; if not, the latter
has much less influence, and that something within
depends very much upon pre-natal influence. You cannot get out of a person emotions which are not embodied
in mental structure, and mental structure depends tre-
mendously upon pre-natal influence. How, can it be
otherwise? The brain of the embryo is the seat of count-
less multitudes of molecular tremors, that are in relation with the actions of the mother's mind and brain, and
it is the sum or outcome of the whole of these intimate,
intricate and impalpable motions which appear in the
illumination of consciousness in the child. It is these
infinitely minute and subtle elements of matter, of the
mother, that minister to and form the mental functions
of the child. The motives that actuate us in the conduct
of life are more often than we suppose, the motives that
actuated our mothers while they were bearing us. A
very supposable conflict in the mind of a young married
woman is the conflict between desires. In the supreme
centres of her brain the desires will fight out their battles,
and, by the struggle which they make for existence,
attain and maintain the equilibrium which will characterize the character of her child in a great measure.
The result of this conflict of desires in the mother will
be partly determined by the native capacity of her mind,
its natural heritages and aptitudes, partly by the degree
and character of the development of her mind, and a great
deal by the earnestness and faith of her prayers to God
for daily guidance, and by her endeavoring to attain to the
true symmetry of life. If we truly want our mental

functions of feeling and knowing, emotion and willing, to be performed in their highest possible manner, a manner approximating to the divine intelligence, then we must ask divine help. Pupils become like their teachers. Those who walked with Plato in the academy learned his wisdom. Those who walked with Aristotle in the lyceum learned his wisdom. The companionship of great men is stimulating to the intellect, and conversation with the divine intelligence is productive of moral developement and expansion. Girls taken away from the environments of vice and crime while young children, and brought up under Christian influences, will, when they bear children, bear those who will be much better in moral qualities than their ancestors were, and this process repeated for a few generations, will stamp out a great deal of vice and crime in the community.

I have to express my thanks to Mr. John T. Cook of the Albany Bar, who has favored me with valuable suggestions, and by correcting the sheets during their progress through the press, has laid me under many obligations.

SUNNYSIDE SANITARIUM FOR DISEASES OF THE NERVOUS SYSTEM.
FLATBUSH, N. Y., *December 1, 1893.*

Medical Jurisprudence of Insanity.

CHAPTER I.

GENERAL CONSIDERATIONS.

THE chief practical issues coming within the range of this volume, which are to be decided by medical witnesses, are :

1. Questions of mental soundness or insanity, as affecting questions of responsibility in criminal cases, and

2. The capacity to make a will or to manage one's affairs.

Our object in this work is to present, as concisely as possible, the application of mental medicine to the purposes of the law. The numerous cases of real or alleged insanity depend, for their final settlement, mainly, if not exclusively, upon medical testimony.

The role of the physician is to point out to the judge and jury that which is disease and that which is not. The whole study of medical jurisprudence is of the greatest practical value both to the lawyer and the physician.

If, in cases determining the validity of a will, or the responsibility of a homicide, the judge and jury were not enlightened by medical testimony, so that law might keep pace with medicine, monstrous wrongs and flagrant injustice would be inflicted in almost every case.

The medical expert, to have his opinions of value, should be a man who has mastered the complex and subtle study of the mental and psychological functions ; he should be a man who thoroughly studies each case that is presented to him by the lawyer ; he should, to be of value in any given case, be a man who, on the witness stand, maintains a calm and dignified composure as a witness, and one who cannot be betrayed into heated rejoinders to counsel, and who does not allow his temper to become ruffled. A medical expert should be summoned on the supposition that he is master of the science of legal medicine. If he is, he will, in his testimony, be neither evasive nor ambiguous, and he will avoid, as far as possible, the use of all technical words or phrases ; neither the court, counsel or the jury can possibly understand technical and scientific language, such as would be appropriate in an address before a medical society. Affectation and pedantry are out of place on the witness stand.

The testimony of experts is necessary for the purpose of arriving at truth in certain medico-legal investigations. Such testimony, viz., that of skilled witnesses, is essential to a due observation and appreciation of facts. Such testimony can only deserve its name and fulfill its function when the witness is really skilled, *i. e.*, when he possesses those qualities of mind, that education of habits, and those stores of information, which alone can make him a competent observer. It is because medical witnesses have often been unskillful in the particular directions in which their evidence has been taken, that so much discrepancy has occurred in their statements. Scientific testimony does not fail in the matter of facts, because it is too minute, too cautious or too true,— rather because it is *wanting* in carefulness, precision and minuteness.

The physician's province is to point out the distinc-

tion between permissible variations within the range of
health, and those departures from the common order
of life, which are inconsistent with the idea of mental
health.

To a physician skilled in psychiatry, nothing appears
more absurd, and nothing could possibly be more in
conflict with the laws which govern mental disease, than
the New York code, which lays down, that if a man
knew the consequences of his conduct, and the difference between right and wrong, he must be held legally
responsible for crime ; yet it happens very often that the
insane are well informed upon these points, and that
sane men are not. The sense of a difference between
right and wrong, in the general or the abstract, is one of
the characteristics of human mental constitution. It
may differ in intensity, in keenness, and in force of
influence upon conduct. It, however, exists as an essential part of our nature, and in some form or another is
present in the most degraded of our species. The application, however, of this sense to particular acts is as
variable as are the conditions of human life, and is the
product of all kinds of influences—the climatic, hereditary, educational and social. If the sense of right and
wrong be destroyed, the individual is less than man ; if
the sense exists, but its application be erroneous, the
individual may be insane, but he may be simply ignorant
or prejudiced. It is a part of our nature to recognize the
distinction in the abstract ; it is not a part of our nature
to determine its particular applications. I have repeatedly seen the insane, not only with a very keen conscience, but actually unhappy with the sense of their
responsibility, while sane men are often met with who
are not troubled in either of these particulars. The
sense is not necessarily absent in the lunatic ; its presence is not a proof of sanity.

The great question in criminal trials is, whether a man
was capable of avoiding the compulsion of disease to
crime ? Could he help it ? Lord Chief Justice Cock-

burn, of England, and Sir James Fitzjames Stephens, in his "Criminal Law," have taken the broad and liberal ground that where there is loss of self-control, caused by insanity, there is irresponsibility. In Stephens' "Criminal Law," 1883, Vol. II, page 130, we find the following, viz.: "Sanity exists when the brain and the nervous system are in such a condition that the mental functions of feeling and knowing, emotion and willing, can be performed in their regular and usual manner. Insanity means a state in which one or more of the above-named functions is performed in an abnormal manner, or not performed at all, by reason of some disease of the brain or nervous system." This is the most liberal definition that ever emanated from the bench, and such a liberal and progressive spirit is very gratifying to see.

Another very important point, and one which Justice Stephens evidently understands, is this : Morbid states of the *emotions* derange the mind, and we not unfrequently see the emotional or affective power of the mind markedly affected while the reasoning powers remain unaffected. There is to-day a decided tendency, on the part of the progressive men of the legal profession, to alter the existing laws, to keep pace with increasing knowledge, while, we regret to say, there is another class of men who have so much professional conservatism that they oppose the law of insanity being brought into reasonable agreement with the knowledge of insanity possessed by physicians.

A strict enforcement of the law will hang many innocent persons.

The more progressive judges all recognize, and from their experience on the bench *know*, that there are forms of mental disease in which, though the patient is quite aware that he is about to do wrong, the will becomes overpowered by the force of the impulse of the mental disease. This is true of very many of the suicidal cases.

A lady whom I saw in consultation a few years ago,

who was a case of suicidal melancholia, told me that she "knew it was wrong to attempt self-destruction ; knew that she should be punished hereafter ; that the thought of it made her very unhappy, as she realized her responsibility ; but she couldn't help the feelings which impelled her to take her own life, or the impulse to do it." Although we cautioned the friends to keep a trained nurse with this patient, and impressed the fact of the insanity of the patient upon the mind of the friends, she eluded their vigilance finally, and took her own life. In such a case, will any lawyer deny that the power of self-control was destroyed by mental disease, and that this was an essential element in the question of responsibility ? Suppose homicide instead of suicide had been the result of the mental disease, would that have altered the fact of the loss of the power of self-control being an essential element in the question of responsibility ?

If, when the law speaks of a person laboring under such a defect of reason as not to know the nature and quality of the act he was doing, etc., it means us to understand *a calm judgment of the circumstances and consequences of the act*, then the judges should so construe it in their charge to the jury, in every criminal case where insanity is alleged by the defense, and counsel for the defense should always request that the judge should so construe the proposition.

If in any case where emotional insanity proper, or reasoning mania, is the type of mental disease which forms the defense, and the district attorney, or the judge, in his charge to the jury, take the ground that the effect of insanity, if any, upon the emotions and the will is not to be taken into account in deciding whether an act done by an insane man did or did not amount to an offense, the counsel for the defense should take the ground, that the proposition that the effect of disease upon the emotions and the will can never, under any circumstances, affect the criminality of the acts of persons so afflicted, is false to every medical truth, and is bad law.

Any law is insufficient and bad which lays down propositions diametrically opposed to science or to medical books of recognized authority. The law of America should be as Sir James Fitzjames Stephens has proposed for England, that *no act is a crime if the person who does it is, at the time when it is done, prevented, either by defective mental power or by any disease affecting his mind, from controlling his own conduct, unless the absence of the power of control has been produced by his own default;* or, as Dr. Bucknill, of England, has suggested as a simplification of Sir James Stephens' bill, "No act is a crime if the person who does it is at the time incapable of not doing it by reason of idiocy or of disease affecting his mind." This is a very needful amendment to our law relating to insanity, and should be in operation in every State of the Union. It is the law, practically, in Pennsylvania. From my experience in medico-legal trials, I consider that there is a pressing need, in cases where the insane prisoner has no friends and no means, for an amendment to our code, which should place rich and poor on the same footing, by providing for an official examination, by mental experts of high standing, into the prisoner's mental state before the trial. If good examiners were appointed, much good would accrue; if incompetent men, the gain to legal medicine would be very problematical. There would be nothing to prevent counsel from calling in the aid of any other experts that they wished, on any given trial; and it would give the friendless insane the services of skilled physicians, who would be remunerated by the county for their services in each case.

We have seen the necessity for expert or scientific evidence. We have also seen that medical testimony may fail of its legitimate effect sometimes if it is incomplete and inaccurate; or, on the other hand, if it is complete, accurate and definite, it may pass beyond the established lines of legal precedent. Our judges are generally men of great attainments, keen appreciation and wonderful habitual fairness. Both judge and jury must often, we should think, be bewildered by the burden of deciding between experts. Their duty, when called upon to weigh expert opinions, is not an enviable one, and they are placed in a very delicate position.

The province of the medical expert in any given case is to represent to the counsel engaging him, the true scientific value of the fact the latter has to deal with. If a medical man be true in his allegiance to science and to his profession, he will never stultify himself in court, and his opinions will soon be regarded alike by court, jury and counsel, as reliable and valuable, as he voices science, instead of appearing as a mere partisan witness.

Honest, unbiased, scientific testimony of an expert, who is assumed to be a scientific man in his chosen specialty, is conducive at once to the good of the individual, the honor of our profession and the cause of truth.

We would insist upon the importance of the study of medical jurisprudence being pursued by both legal students in the law schools and by medical students in the medical colleges, under competent professors.

Medically, insanity is a disease of the body, affecting the mind by deranging its faculties, and causing such suspension or impairment of the healthy intellect, the emotions or the will, as to render an individual irresponsible.

It might also be defined as a diseased state of mind, due to ill health, accompanied by more or less absence of self-control, and impairment, in a marked degree, of the intellect, the emotions or the will; and showing itself psychically by depression, exaltation or mental weakness, and by disorders of sensation, perception or conception. We regard the former as the better medico-legal definition of the two.

When a criminal case is presented to the lawyer, if mental disease be suspected, a physician is consulted to examine the prisoner, give his opinion to the counsel,

and if favorable to the latter's view of the case, to testify
as to the prisoner's irresponsibility. These expert witnesses should form their judgments with the greatest
care, and then express themselves in the plainest terms.
We cannot over-estimate the importance of having clear
ideas and of expressing them clearly ; if the expert witness does this, he will be successful. Want of accuracy
of thought, and of distinctness of expression, mars many
an expert's opinions given in conrt. It should be the
duty of the expert to examine the prisoner sufficiently
often to thoroughly satisfy himself as to the existence
of mental disease. The case for the defense may be considered especially strong when the insanity of the prisoner can be proved to be hereditary, when there have
been previous attacks, or when epilepsy is present. In
the case of the People v. Nelly Vanderhoof, the defendant
was charged with killing her newly-born babe. Careful
examination into the circumstances of the case revealed
the fact that the prisoner had suffered from epilepsy from
birth ; that the family were saturated with the disease.
Nelly Vanderhoof was a young unmarried woman,
suffering from the strong moral shock of seduction and
desertion, and the irritable condition of the nervous
system produced by epilepsy ; she also had, when we
first saw her, a considerable degree of uterine derangement. When the writer examined her at the Tombs, she
had apparently no realizing sense of the enormity of her
crime, and the mental tone had become very obviously
impaired as the result of epilepsy. After investigating
her mental condition, we reported to her counsel that she
was, in our opinion, irresponsible, and that epilepsy was
the phase of mental disturbance that prompted the
criminal act. During her past life, she had been many
times under the dominion of that blind fury, so frequently exhibited by epileptics immediately before or
after a fit. Her mind was generally so impaired that she
was seemingly incapable of controlling the feeblest
impulses of passion ; she was laboring under a disease

which almost invariably impairs the mind ; she had a sister demented, as a result of the same disease, a resident of one of the New York institutions for the insane; her father was a case of dipsomania ; her mother had twice attempted suicide ; such was the prisoner's mental condition and her family history. The trial took place before Judge Van Brunt, in the New York Oyer and Terminer, April 9th and 10th, 1885. The People were represented by Assistant District Attorney Fellows, who, in trying this case, deserved great credit for his enlightened and humane views respecting the exculpatory effects of this disease. The prisoner was acquitted, the jury rendering a verdict of "not guilty," on the ground of insanity. Judge Van Brunt delivered a very fair, impartial charge, acknowledging the exculpatory effects of epilepsy.

LEGAL RELATIONS OF EPILEPSY.

Not unfrequently the criminal lawyer will become engaged in cases in which epilepsy is the phase of mental disturbance that prompts the criminal act. Upon careful investigation he will generally be able to find epilepsy or insanity existing either in the parents or grandparents of the prisoner. Epilepsy is sufficient alone to produce complete irresponsibility. The mental powers become impaired as the result of epilepsy, and epileptics have the irritable condition of the nervous system produced by this disease. Such persons are prone to be under the dominion of that blind fury generated by the disease, both before, after and between the fits. The mind of epileptics is often so impaired that they are seemingly incapable of controlling the feeblest impulses of passion. Epileptics labor under a disease which *almost invariably* impairs the mind. The brain and nervous system of these persons is apt to be in such a condition that the mental functions of *feeling* and *knowing*, *emotion* and *willing*, are not performed in their regular and usual manner. One or more of the above-named functions is performed in an abnormal manner, or not performed at all. The outbursts of maniacal fury and destruction and homicidal impulses of epileptics are peculiar, in that the duration of the morbid state is short and its cessation sudden. There is no well-educated physician in any country who does not know that the disease of epilepsy produces a modified responsibility in all the subjects of said disease. In a large number of cases the actual or comparative sanity of patients, for considerable intervals of time, the freedom from irascibility, passion or violence, when removed from circumstances calculated to irritate, render it difficult to place such persons under restraint until an overt act has been committed which necessitates sequestration.

Very often the character of the mental disturbance, the paroxysmal gust of passion, the blind fury without an adequate cause, indicate the presence of epileptic insanity, and take the place of epileptic fits. Masked epilepsy is indicated by eccentric acts or a sudden paroxysm of violence without a distinct epileptic seizure.

Unmistakable epileptic fits occur at one period of a patient's life, while at another, maniacal symptoms take their place. When mental symptoms appear to take the place of a fit, there is a transitory epileptic paroxysm. All acts, soon after epileptic fits, are automatic, and the patient is irresponsible.

Elaborate and complex actions may be performed while a patient is unconscious. In different cases there are different degrees of recollection. As in other forms of insanity, there may be a motive mixed up with an insane condition.

There may be a motive and calculation in some cases, which, in some rare cases, control the misdeeds of epileptics. It is certain that the victim of a disease which takes away from him all control over himself, even when he remains capable of distinguishing between good and evil, cannot be held responsible for acts which he accomplishes without will, and in an automatic, and therefore unconscious, manner. There is no epilepsy without unconsciousness. Epileptic seizures vary in severity from a simple vertigo, scarcely discernible by others, to the most violent convulsive fit, lasting from five minutes to some hours. Anger, fright, or any strong moral emotion, is very liable to produce a paroxysm. Epilepsy tends almost invariably to destroy the natural soundness of mind. A direct, though temporary, effect of the epileptic fit is to leave the mind in a morbidly irritable condition, in which the slightest provocation will derange it entirely. This was precisely the state in which Lucille Yseult Dudley was in when she shot O'Donovan Rossa. She had, within a few days, had nineteen epileptic fits, and the provocation was, the news which had arrived from London of the dynamite outrage, of which she imagined Rossa to be the direct instigator. Her criminal act was the result of the morbid irritability which succeeded the epileptic paroxysm. In epileptics, it is not uncommon to observe attacks of mania which are often characterized by a high degree of blind fury and ferocity. During the attack the patient is unconscious, so that his acts, whatever may be their nature, cannot make him liable to legal punishment. The passionate impulse to kill, in masked epilepsy, is substituted for ordinary epileptic convulsions. Instead of a convusion of muscles, the patient is seized with a convulsion of ideas. An epileptic convulsion may not occur, but may be represented by sadness, dejection, by sullenness, by ebulitions of rage and ferocity, a *mania transitoria*, signalized by suicide, homicide and every modification of blind and destructive impulse. The awakening from epileptic stupor may often resolve itself into an outburst of mental derangement, manifested by extreme vehemence, violence and destructiveness. A crime resulting from epileptic psychical phenomena may be accomplished with comparative deliberation, and, as we have before remarked, there may be a motive mixed up with an insane condition. All epileptics are impressionable and excitable, and epileptic attacks are often replaced by irresistible, homicidal tendencies.

A patient may recognize his impulses as illegal, but irresistible. In epilepsy, dreamy, mental states and imperative acts appear and disappear with great suddenness. If an epileptic who is a prisoner, having committed some overt act, has premeditated the act, that does not prove that the said prisoner was not insane, or that he could control his insane desire. On the contrary, it might be a still stronger proof of his insanity, that under the circumstances in which he was placed, he would do an act from the fearful consequences of which it would be impossible for him to escape. Every day there are examples in insane asylums of insane persons committing crimes that they have premeditated. Premeditation is no proof of a prisoner's sanity. Epileptics who commit overt acts are very frequently indeed not in a condition to realize the nature and quality of the act they are doing, or to know that the act is wrong. Homicide or assault, with intent to kill, is not criminal, in our opinion, if the person by whom it was committed is, at the time when he commits it, prevented, by any disease affecting his mind, from controlling his own conduct. If any person, at the time of committing an overt act, is suffering from incapacitating weakness or derangement of mind, produced by disease, then they are insane and irresponsible. It is very seldom that such facts cannot be elicited if they are present, and trials to-day are seldom unfair. Of course there are painful exceptions, where public prejudice virtually tries and decides a case, but this seldom occurs.

It should be distinctly understood that it is a scientific fact that if an epileptic or a maniac, subject to delusions, conceives a desire to murder, that he will be as incapable of resisting that desire, as he has already proved himself incapable of resisting either his fits or his delusions. Delusions of the insane defy the evidence of their own senses, the efforts of their reason, the testimony of their

sane neighbors, and the remonstrances of their friends ;
and their impulses always have, and always will, prove
just as irresistible, when confronted with their knowledge
of the distinction between right and wrong and the
remonstrances of their consciences. Mental disease does
not deprive a man necessarily of the knowledge and consciousness of the law. It is inhuman, unscientific and
diametrically opposed to every known psychological law,
to only hold the insane man irresponsible for his act if
his mind can be shown to be so unconscious of right and
wrong that he is incapable of appreciating the law and
its requirements.

The law to-day, in New York State at least, insists
upon a test of insanity which every physician of experience, or whose opinions are of any value respecting
insanity, says it is impossible to apply.

The jury take their oaths that they will try a given
case fairly and impartially upon the evidence ; that they
can do it without bias or prejudice on account of any
opinion which they have formed ; that they will try the
given criminal case without being affected or influenced
on account of any circumstances which surround the
criminal transaction ; that they will try the case according to the sworn testimony of the witnesses, and that
they have no opinion of the law which shall govern said
case. It is rarely, in a great case, that each of the
gentlemen, before entering the jury box, has not read
accounts of the affair, from which he has formed some
impression in reference to the criminal transaction.
Before, however, they enter the jury box, they have to
state, on their oath, that they believe they can lay aside
their previously formed opinion, that they can enter
the jury box, listen to the evidence, and determine
the facts anew, according to law and the evidence, without being influenced by any previously formed opinion.
This duty devolves upon each juryman, and it is a duty
he owes to the public, to the prisoner, and to his own
conscience. The jury should not, on going to the jury

room, enter into any hasty or passionate discussion of the questions involved, but coolly and calmly reason one with another, to the end, if possible, that they may bring their minds to a common conclusion, and in so doing, determine the right in any and every case.

The law in New York, bearing upon the question of insanity, is as follows: *"A person is not excused from criminal liability as an idiot, imbecile, lunatic or insane person, except upon proof, that at the time of committing the alleged criminal act, he was laboring under such a defect of reason as either not to know the nature and quality of the act he was doing, or not to know that the act was wrong."*

Medically speaking, the law errs in making the test of responsibility the capacity of the person to distinguish between right and wrong at the time of and in respect to the act complained of.

The question, according to the present defective law, is, Was the prisoner, at the time of committing an overt act, in such a state of mind as to know that the deed was unlawful and morally wrong? If he was, then he is responsible. If he was not, then he is not responsible.

The law bearing upon the question of insanity should be codified and amended, and the question should be, Was the prisoner's brain and nervous system in such a condition that the mental functions of feeling and knowing, emotion and willing, could be performed in their regular and usual manner? Was the man capable of avoiding the compulsion of disease to crime? Could he help it? Was the prisoner prevented, either by defective mental power or by any disease affecting his mind, from controlling his own conduct? The law should take the broad and liberal ground that where there is loss of self-control, caused by insanity, there is irresponsibility. When this is done, then, and only then, will the law of insanity be brought into reasonable agreement with the knowledge possessed by physicians.

Every case is to be judged, not by any ordinary standard, but by the change in the person himself. Everyone, therefore, becomes the measure of himself, and we are to inquire what the individual was, and what he has become, through disordered conditions of the brain. A medico-legal point of great importance, which cannot be too strongly insisted upon by lawyers in every criminal case where insanity is alleged, is this: That the instability of nerve element implied in heredity, has a positive influence and is a definite power. It is an important point to bring out in some cases, that a man may be in a condition bordering on insanity, and by exciting causes be drifted over to the insanity side.

On the question of change of character in a person accused of overt acts, and whose insanity is alleged, I would call the attention of the legal profession to the statement in Bucknill's "Essay on Lunacy," page 33: "A change, therefore, with impairment or perturbation of function is the chief test of cerebro-mental disease. It may take the same direction as the original character; and persons naturally timid or daring, cautious or reckless, generous or selfish, may have their natural bias of mind quickly developed in excess; or the change may reverse the character, and the patient may exhibit a striking contrast to his former self, or may take some strange direction which no one could guess at beforehand. Nothing can appear more wayward and uncertain than the direction which insanity takes in its development." That the insane act from motives, as the sane do, and that they are moved by fear, revenge, hatred and jealousy, is well illustrated in the case of Renshaw, who, entertaining a feeling of bitterness against Dr. Gray, Superintendent of the Insane Asylum at Utica, armed himself with four pistols, several pounds of cartridges and a bowie-knife, put on his feet rubber boots, that he might make no noise, and stole noiselessly along the hall to Dr. Gray's office, deliberately discharged his pistol at the doctor's head, the ball penetrating his face, and turned and fled. In a short time he went volun-

tarily to the jail and delivered himself up. The posses
sion or sight of a deadly weapon often suggests to the
insane the commission of an act of violence.

The lawyer should be equally instructed with the
physician, as to what sort of an examination his client,
if insanity is advanced as a plea in a criminal case,
should have, in order that the fact of mental unsoundness may be elicited, if it exists. There are what physicians call premonitory symptoms of mental unsoundness.
There is altered health, altered or perverted sensations,
in some cases loss of muscular power, sleeplessness very
frequently, excessive irritability, alterations of temper,
excitability, tendency to laugh or cry, suspiciousness
without adequate cause, unreasonable likes and dislikes,
sometimes intense egotism, loss of memory, confusion of
ideas, inability to think, write or speak connectedly,
alteration in manner of speaking, and other changes in
the intellect, emotions or behavior.

RULES FOR THE EXAMINATION OF PERSONS SUPPOSED
TO BE OF UNSOUND MIND.

Every lawyer of experience knows that in medico-legal
trials the physician who is to examine a person in whose
defense the plea of insanity is to interpose, cannot be too
careful in his examination. He should make a written
examination, and should, when he gets home, make a copy
of it for the lawyer who is to defend the case.

First. He should observe the general appearance and
the shape of the head; the complexion and expression of
countenance; the conformation of the body; the gait and
movements, and the speech.

Second. Ascertain the state of the general health, of
the appetite and digestion, of the bowels, of the tongue,
skin and pulse. Note especially the presence or absence
of febrile symptoms, as an important aid in distinguishing delirium from madness. Ascertain whether there is
sadness or excitement, restlessness or stillness, and
whether the sleep is sound and continuous or disturbed

and broken. In females, the state of the menstrual functions should be inquired into.

Third. The family history should be traced out, in order to ascertain whether there is any hereditary predisposition to insanity, whether any members of the family have been subject to fits, or have betrayed marked eccentricity of behavior.

Fourth. The personal history should be ascertained with equal care. If the mind appears unsound, ascertain whether the unsoundness dates from birth, or from infancy, or from what time. If the unsoundness has supervened later in life, whether it followed any severe bodily illness, accident, mental shock, long-continued anxiety of mind, repeated epileptic fits, or course of inebriety.

Fifth. Inquire whether the present state of mind differs materially from that which existed when it was reported to be sound ; and whether the feelings, affections and domestic habits have undergone any marked change.

Sixth. Ascertain whether the existing unsoundness is a first attack, and if so, whether it began with depression or excitement. Did it follow a period of melancholy, pass into mania, and then into slow convalescence ? Has the patient suffered from epilepsy ? If any signs of general paralysis are present in the speech or gait ? Has the patient squandered his money, grown restless and wandered about, exposed his person, committed petty thefts, or had illusions of wealth or grandeur ?

Seventh. If the physician desires to test the capacity of the mind, it must be tested by conversation directed to such matters as age, birth place, profession or occupation of parents, number of brothers or sisters and near relations ; common events, remote and recent ; the year, the month or the day of the week ; the name of the municipal Mayor, the Governor of the State, and the President ; and of persons best known and talked about.

2

The power of performing simple operations in arithmetic, and the knowledge of the value of money, should be tested, and the power of repeating simple forms of words in general use, such as the "Lord's Prayer," etc. In testing the power of attention, merely negative or affirmative answers to leading questions, should be distinguished from such replies as indicate judgment and reflection. If the inquiry relate, not to the capacity of the mind, but to its soundness in other respects, delusions should be sought for by conversation directed to those topics that are most likely to interest and excite the mind. The state of the moral feelings will be tested by conversation directed to relatives and friends. In cases of psycho-sensory insanity, diligent inquiry should be made into the motives which might have led to the commission of the act of which the party was accused.

Eighth. The physician should insist in full opportunity being given him of forming his opinion. He should not usually content himself with a single visit. In cases of great difficulty, he should insist that the party be placed for some time under his observation.

Ninth. When undergoing examination in court, the medical witness is recommended generally to avoid definitions of insanity, on the plea that mental, like bodily, diseases can be described better than defined.

Respecting some special forms of mental alienation, we desire to express the decided opinion that kleptomania, erotomania, pyromania, dipsomania, and suicidal and homicidal mania, are all distinct varieties of insanity. Kleptomania is most common in women, placed by their wealth beyond the reach of vulgar temptation. Cases of theft are also often met with in epileptics ; they care not what the value of the article is. Erotomania is an example of one of the strong impulses of our nature that is sometimes placed, by morbid excitement, beyond the restraint of reason and conscience. Pyromania is most frequent in young girls subject to menstrual disturbances. Dipsomania is a well-recognized form of mental unsoundness, and we would strongly maintain the necessary dependence of suicide on insanity. The majority of cases of homicidal mania, in our experience, have been among women, and are the result of grief, anxiety, from uterine disease, at the menstrual period, at the climacteric period, and often at delivery, especially when complicated with seduction and desertion. Women at these times are in a peculiarly nervous state, not unfrequently, I am led to believe, accompanied by impulses to crime, and we do not consider them as responsible for overt acts committed at such times, especially when an overt act is opposed to the whole previous character of the woman.

CHAPTER II.

MORBID SEXUAL PERVERSIONS AS RELATED TO INSANITY.

RELIABLE facts are of course most difficult to obtain, and such figures reveal little of the real truth; the extensive mental mischief done; of which there can be no doubt whatever. These morbid sexual perversions are most commonly met with in a love of the same sex in both male and female subjects. They are generally associated with more or less mental weakness and a state of psycho-sensory insanity, and are the results of a faulty nervous organization. Although sexual perversion may not necessarily be by itself a perfect proof of insanity, yet in any given case where we find a female developing sexual love for one of her own sex, or a male subject, from childhood up, showing feminine tendencies, shunning boyish sports, assuming female costume, and developing platonic and sexual love for persons of the same sex, we may strongly suspect the existence of psycho-sensory insanity; by which I mean an abnormal state, in which there is a morbid perversion of the natural feelings, affections, inclinations, habits, moral disposition, and natural impulses, without any remarkable disorder or defect of the intellect or knowing or reasoning faculties, and without delusions. It seems to be a reasoning monomania and sometimes an erotomania. The conduct is affected more than the conversation, but the patient is none the less insane. We often find the dispositions and habits changed, the affections perverted, and, finally,

either a maniacal excitement, during which overt acts of
a destructive character are committed, or else a weakening of the mental faculties, ending in dementia. We
may have an instinctive psycho-sensory insanity, hurry-
ing the patient on to instinctive and automatic acts not
preceded by reasoning, and a reasoning psycho-sensory
insanity, determining acts which are the consequence of a
certain intellectual operation. That a person premedi-
tates a crime is no proof at all that such a person is not
insane, as the insane premeditate very often the overt
acts they commit, and where the compulsion of disease
toward crime is so strong that the patient's will is weakened by the *vis-a-tergo* of the insanity.

There is the absence of self-control produced by disease, and the patient, though well aware that the act is
wrong in some cases, has no power of resistance at all.
The mental functions of feeling, knowing, emotion and
willing are not performed in their regular and usual
manner, and there is very often morbid delusive conception or perception of subjective origin, causing change of
mental character as compared with former self or normal
ancestral type, through organic conditions originating in
disease within the system, external motives playing but
a secondary part when they influence at all the mental
conduct. Change of character is the ultimate symptomatic expression of insanity, change of mental conduct
the immediate. The condition of mind is not voluntary ;
it is the product of disease. The most striking features
of insanity in general, and the strongest proof of the
presence of any of its forms, is the change for the worse
that takes place in an individual's character and habits.
Of course, when we have to deal with a congenital deficiency, the natural character itself being abnormal, this
test does not apply. There is very often a true congenital moral deprivation, with strong animal propensities,
which makes a person practically insane from birth. In
these cases there is no sense of shame or remorse.
Psycho-sensory insanity occurs frequently in early life.

The intellectual faculties appear unimpaired. Both males and females appear to perfectly perceive, and know, and judge. There is no delusion, yet they are insane, and as much need medical care and restraint as the worst forms of mania. There is an entire perversion of the moral principle, and there are no good or honest sentiments. They are actuated by their impulses and by the most depraved motives, but it is disease, not crime, that they are suffering from. Many such cases, from being refined and virtuous, become coarse, depraved, licentious, dishonest and reckless. Some of them are incorrigible thieves ; others exhibit morbid sexual per-version, manifested by love of the same sex, sometimes intense and pure, but more often intense and most impure. Sometimes we find masked epilepsy in these cases, destructive acts and gusts of passion taking the place of a well-marked fit. They are cases of *petit mal*, and are very dangerous. If such cases could be carefully studied, as they are not, as they rarely come under observation until they have committed some overt act, it would be found that a great many persons, perverse and capricious and depraved throughout their entire life, are really cases of psycho-sensory insanity.

In cases of morbid sexual perversions we sometimes find a true erotomania, characterized by excessive love for an object, a mental affection, in which amatory delu-sions rule the patient, as religious delusions rule in theomania. It is different from nymphomania and satyriasis, in that, in the latter affections, the disease originates in the organs of reproduction, a constant stream of irritative impressions being sent practically without cessation to the brain. In erotomania the seat of the disease is in the brain itself. The two sometimes, however, co-exist, and patients will often pass far beyond the limits of propriety where we can find no trouble in the reproductive organs. I have in mind such a case at present, where, upon the advice of a celebrated gynæco-logist, the ovaries were removed and were found to be perfectly healthy, but the patient experienced no relief. After death, when the brain was examined, there was found an extensive area of what had been irritation and inflammation, followed by hardening of the most pronounced nature. In this case there had been many attacks of hystero-epilepsy. In erotomania there is depression of mind and body, emaciation, and, if a cure is not accomplished, the patient rapidly sinks and dies. We very often find erotomania following a religious melancholia, and it occurs more frequently in females than in males.

There are many women with perverted sexual instincts and a psycho-sensory insanity, who, at each menstrual epoch, become possessed with a strong homicidal impulse, and those nearest and dearest are often the ones to suffer death at their hands, or perhaps any one who may at the time displease them. Revenge and other motives are not unfrequently mixed up in insanity with such symptoms. Usually, we think, there will be some decided evidence of heredity, or of a change in the feelings prior to the commital of the overt act, but there are many cases, we think, in which the act, and the act alone, may constitute sufficient evidence of the insanity of the homicide. There are often active organic influences of a morbid nature, which, though not externally noticeable, may, when disturbed and disordered at the moment of action, impel the person toward crime. In these cases I think there is generally a *petit mal*, and this disease often weakens volition without any external mental symptoms. In doubtful cases it is very important to make the most searching inquiries as to whether epileptic seizures, however slight and transient, have been noticed. Sometimes a transient dizziness, or pallor, or a momentary mental blank, is the only indication of the existence of a masked epilepsy, and very frequently patients are only under the influence of destructive impulses when such an attack is threatened, and an overt act often takes the place of a fit in such persons.

I have a patient under medical observation and treatment at the present time, who is highly educated, refined,
and virtuous, and apparently normal in every respect,
with not a trace of a delusion, and who is even morbidly
sensitive as regards right and wrong ; who suffers from
epileptic vertigo, which generally comes on either before
or after the menstrual epoch, and sometimes during the
epoch ; who has had strong compulsions of disease
toward crime in the form of suicide, and who has been
tortured by the thought that she might some day give
way to it. It passes away almost immediately, and she
is cheerful and sunshiny in the intervals between the
paroxysms, and is a devoted wife and daughter. It must
be a very simple thing for a homicidal to take the place
of a suicidal impulse, and, if she should ever in the
future give way to such an impulse, the public and her
friends would be greatly astonished, and perhaps refuse
to believe her irresponsible for it, as they have never
seen any indication of mental mischief. Every physician, however, can see that the question here should be,
not could she distinguish between right and wrong at the
time of the commission of the overt act, but, could she
help it ? Could she avoid the compulsion of disease
toward crime ? Was the act the outcome and product of
the epilepsy ? No act which is the product of disease of
the body affecting the mind by deranging its functions
and causing a suspension or impairment of the healthy
intellect, the emotions, or the will, can be construed into
a crime, and physicians should always voice this in the
court-room when they have to give their professional
opinion. It is time that the laws should be amended to
keep pace with medical science, and the absurd right and
wrong test of insanity, which is no test at all, as every
progressive physician knows, should be relegated to the
dark ages, where it properly belongs. Two-thirds of the
insane know the difference between right and wrong as
well as the sane do, but there is the absence of self-control produced by disease of the brain. Shall we strangle

sick people to death in obedience to traditional dogma?
It is time for the medical profession to come to the front
and voice science in this matter. A scientific truth is
never a dangerous doctrine, and we are bound to go
where science leads us if we are true followers of the
noble profession of medicine.

There are, unquestionably, instinctive monomanias,
free from all complications. Associated with pathological sexual perversions we find the abnormal mental ideas,
born of insanity, tyrannizing over the patient's thoughts
and acts, and a psycho-sensory insanity—an insanity of
conduct, feeling, or impulse, or all combined, without
such appreciable intellectual derangement that it would
be recognized as insanity without the display of morbid
feeling, impulse, or conduct.

The great diagnostic mark in these cases is the
predominance and overshadowing and overmastering
character of the aberration of the moral faculties over
the faculties of the understanding. It is seen in action
and conduct, rather than in words. There are morbid
changes in the appetites, propensities, and feelings.
These cases are less understood and studied than other
victims of mental disease, because, as we have said, they
are rarely brought into notice until an overt act is committed by such a person. Imperative conceptions and
morbid impulsions are very characteristic of this class of
cases. The morbid mental condition of these cases of
psycho-sensory insanity is a basis fact in psychiatric
symptomatology which cannot be reasoned away. Delusion is comparatively exceptional, while perverted feeling
and conduct is never absent. The wishes, inclinations,
attachments, likings, and dislikings, are morbidly
changed, and this change appears to be the origin or to
lie at the very foundation of any disturbance which the
understanding itself may seem to have sustained, and
even in many instances to form thoughout the sole
manifestation of the disease of psycho-sensory insanity.
In one instance I have known of this morbid sexual love

for a person of the same sex, starting, probably, with
some one girl, of a faulty nervous organization, in a young
ladies' seminary—almost assuming the form of an epidemic (genesic erethism)—and several young ladies were
brought up before the faculty, and were told that sum-
mary dismissal would follow if this were not at once
dropped. The terrible mischief which was thus arrested,
and doubtless originated with an insane girl, in this case
evidently assumed an hysterical tendency in others not
insane, but who might have easily become so if they
were neuropathically endowed, as they doubtless were.
Sometimes, in cases of masturbation, perverted sexual
feelings, such as forming morbid attachments for persons
of the same sex, are quite marked. Dementia and death
is generally the end of these cases, unless the general
health is improved and the weakened will-power
strengthened. A nervous temperament, stimulating
diet, improper associations and training, obscene literature, an arrested cerebral development, partial phymosis,
with hyper-æsthesia of the glans penis, are some of the
most frequent causes of this sexual vice. Masturbation
is an exciting cause of insanity; the general health of
insane masturbators is always impaired; the diagnosis
may be difficult at first, but is easy after the first stage ;
the prognosis is very unfavorable unless the practice is
stopped ; daily exercise to the point of fatigue is essential
to treatment, and tinct. gentian comp. seems to be the
most valuable tonic to employ, with a nutritious diet,
but no stimulants—tobacco or coffee—and work of some
kind is a necessity for these patients. It may be noticed
that a genesic erethism may reign as an epidemic at times.
It did so in antiquity, in the Middle Ages, and in modern
times. Julius Cæsar, Augustus, Tiberius, Caligula,
Claudius, Nero, Galba, Otto, Vitellius, Titus, Domitian,
Eliogabulus, Trajan, Adrian, and Commodus were given
over to ferocious and brutal sensuality, which was due to
their hereditary organization. Likewise Agrippina Mes-
salina, Poppæa, Domitzia, Sœmis, the two Faustinæ,

Crispina, Titiana, the two Juliæ, Noua, Celsia, and Lucillæ, Roman empresses, all corrupt women, with aberrations of the genital sense.

In the Middle Ages there was the belief in demons transformed into men for the service of women, and demons transformed into women for the service of men. There were neuropathic epidemics of every sort. Maria of Aragon, Joanna of Naples, Sextus IV, Julius III, Francis I, Henry III, Henry IV, and Louis XIV, all showed morbid sexual perversions. In modern times we find the same thing during the regency and reign of Louis XV, commencing with Philip of Orleans. The Princess Elizabeth, daughter of the Regent, Queen of Spain, the Count de Charlais, and the Marquis de Sode, all abandoned themselves to abnormal sexual perversions and scandals. Some of these cases probably belonged to atavism and some to pathology. Diseases of the genital organs may produce masturbation, nymphomania and satyriasis, but in general pathological sexual perversions we think the brain and nervous system primitively and principally affected, and these psychical disorders are sometimes incurable. Heredity is a strong cause, and there is a correlation of morbific force in disease which may give us a transformation in heredity. Whether the median lobe of the cerebellum is at fault in these cases, as has been asserted by Valentin, Wagner, Susanna, and others, I do not pretend to say, and I do not think we know. We find in the insanities in which the sexual functions are concerned, intellectual anomalies, insanity of puberty, post-connubial insanity, insanity connected with the menstrual period, with the menopause, erotomania, nymphomania, satyriasis the psycho-sensory insanity with pathological sexual perversions, and finally, violation or rape.

There are certainly pathological perversions of the sexual sense in which these passions assume a morbid character and give rise to a true partial delirium, limited to the genital sense and sparing the integrity of the other

faculties of the mind.   Menstrual disturbances are a
fruitful cause of psychical abnormalities and also ovarian
affections.   Many homicides and suicides owe their origin
to erotic conditions and erotic delirium, while nymphomania transforms the most modest woman into the most
degraded one.   I believe that, in all cases of pathological
sexual perversion, science could, if it had the opportunity, detect profound alterations in the brain or in other
parts of the human body.   It is very important to
ascertain in all these cases, if possible, if there has been
freedom of the will, or whether disease has produced a
compulsion toward crime which the will was powerless to
restrain, and I think it very possible to do this.   We
certainly find apparently sound reasoning power with the
most profound pathological sexual perversions in many
cases—a truly psycho-sensory insanity—an insanity of
conduct, feeling or impulse, or all combined.   When
these insanities have a purely physical origin the prognosis is favorable ; when the origin is psychical it is
unfavorable, and where the origin is at once psychical
and physical the case is almost incurable.   If very careful
search be made, I think it not uncommon to find slight
disorders of the intellect, which would, however, attract
no notice at all from those not skilled in psychiatry.   It
is very important to distinguish psychic atavism, which
is the sudden return of the most remote psychic characteristics in men and women of the highest races, from
cases of psychological aberration.   That there is this
regressive phenomenon of thought or feeling, or of both
these momenta of nervous life, there is not the least
question.   It is not always easy to make a clear and
precise distinction between a psycho-pathological phenomenon and an analogous one of regressive atavism ;
frequently it is very difficult, and in some rare cases it is
impossible, the phenomena being identical in form,
degree, result, and permanence.   The pathological phenomenon, much as it may resemble the atavic phenomenon, is essentially different from it.   The criminal may

be a sick man or a man in the most robust health, and
his crime may belong either to pathology or regressive
atavism. We may have a criminal monstrosity, a type
of atavism and a psychic monstrosity showing repulsive
acts of crime and moral degradation, but not a true
criminal at all. The force of inertia may keep psychic
phenomena from appearing for an indefinite time, but
psychic atavism discloses them sooner or later. To lie
concealed does not mean to be destroyed. Christian
civilization has taught the cerebral moderating centres to
hide the genital, the cruel and the filthy atavism, but
this appears when these centres cease to act, or the automatism of old and latent force succeeds in overcoming it.
An insane man becomes a murderer (like his ancestors,
when primary man lived, defending himself against
animals and men, and hunting and fighting was his principal occupation) because atrophy or degeneration of the
moderating centres annuls suddenly the whole progressive evolution of civilization ; on the other hand, a sane
man kills his fellow-man, oppressed by an intense hatred,
which, by its extraordinary power, silences the action of
the moderating nerve-cells. This example enables us to
distinguish, in a measure, between atavism and pathology.
The first of the two murderers, being diseased in mind,
belongs to pathology, and is not responsible, because
there is loss of self-control produced by disease and
his act is the product of disease. The second is perfectly healthy, and belongs to normal psychology, and
hence to man's tribunal, offering us a fact of psychic
atavism. The close relation between cruelty and lust,
which I have mentioned in some of the Roman emperors
and empresses, also forms part of the history of psychic
atavism. Many sexual perversions have in them atavic
influences, often very difficult to separate from the
pathological element, but I think that this can generally
be done by very careful investigation, and it is very
important not to confuse atavism with pathology in
medico-legal study and investigation.

Sexual influences, in connection with insanity and crime, are not sufficiently studied. In many of these cases there is a strong and sudden revulsion of feeling, in which love and confidence are succeeded by the deadliest hate. There is more or less mental disturbance exhibited, not so much in the form of delusion as in that of paroxysmal fury and uncontrollable criminal impulse. In these cases there seems to be an entire abandonment of every interest and feeling not connected with the single purpose of revenge. The person gives herself up to justice, glories in the bloody deed, and is careless of the future. An overt act, not to be distinguished at first, perhaps, from the ordinary criminal deed, is often prompted more by these physiological movements, characteristic of the female constitution, than by well-considered motives, or strong, healthy feeling. With women it is but a step from extreme nervous susceptibility to downright hysteria, and from that to overt insanity. In the sexual evolution, in pregnancy, in the parturient period, in lactation, strange thoughts, extraordinary feelings, unseasonable appetites, and criminal impulses may haunt a mind at other times innocent and pure. We must never ignore the presence of the sexual element in the phenomena displayed by this class of cases, as nervous erethism, excited even by courtship, has a controlling influence over the female will. The common reluctance to attribute insanity to this class of persons arises principally from the fact that they act from a rational motive—revenge—but this is not at all incompatible with insanity, as the insane often act from rational motives, and premeditation and revenge are met with very frequently in the insane. We must also bear in mind that these cases are often persons of a naturally irritable and nervous temperament; a neuropathic constitution needing but a slight exciting cause to induce insanity. The Alice Mitchell case in Memphis, Tenn., is a very good example of this class of cases.

CHAPTER III.*

AMOUNT OF EVIDENCE NECESSARY FOR ACQUITTAL
WHERE DEFENSE IS MENTAL DISEASE.

THERE are three doctrines presented by the courts as to the amount of evidence necessary for an acquittal where insanity is the defense.

In Ohio, Minnesota and Delaware :

1. That as presumption of innocence attends the defendant on trial, and presumption of sanity attends the case of the State, the same amount of evidence is necessary to remove the one as the other. And since the State must establish the defendant's guilt beyond a reasonable doubt, the defendant, when he pleads insanity as a defense, must establish that beyond a reasonable doubt. Clark v. State, 12 Ohio, 495 ; Bonfanti v. State, 2 Minn., 123; State v. Pratt, 1 Houston Crim. Cases (Del.), 249.

In Maine, Massachusetts, Ohio, West Virginia, Virginia, Louisiana, Missouri, California, Iowa, Arkansas, Pennsylvania, New York (?), Kentucky, Alabama (?), Texas :

2. That insanity is a simple question of fact, to be proven like any other fact ; and any evidence which reasonably satisfies the jury that the accused was insane at the time the act was committed, should be deemed suf-

---

* See Brown's Criminal Law, 7–8 ; 17 Am. L. Rev., 921 ; 4 Crim. L. Mag., 509 ; 14 Cent. L. J., 2, 21.

ficient for an acquittal. State v. Lawrence, 57 Me.,
574; Com. v. Rogers, 7 Metc., 500; 41 Am. Dec., 458;
Bond v. State, 23 Ohio St., 349; State v. Strauder, 11
W. Va., 747; Baccigalupo v. Com., 33 Gratt. (Va.),
807; State v. Vann, 82 N. C., 631; State v. Coleman, 27
La. Ann., 691; State v. Redemeier, 71 Mo., 173; People
v. M'Donell, 47 Cal., 134; State v. Stickley, 41 Iowa,
232; McKenzie v. State, 26 Ark., 334; Pannell v. Com.,
86 Pa. St., 260; O'Connell v. People, 87 N. Y., 377;
41 Am. Rep., 379; Kriel v. Com., 5 Bush., 362; Boswell
v. State, 63 Ala., 307; 35 Am. Rep., 20; Webb v. State,
9 Tex. App., 490.

In New Hampshire, Michigan, Illinois, Alabama (?),
Kansas, Indiana, Nebraska, Georgia, New York:

3. That when a person is accused of the commission of
a crime and pleads that he was insane at the time, evidence of sufficient weight to raise in the minds of the
jury a reasonable doubt of defendant's sanity at the time
the act was committed entitles him to an acquittal.
State v. Bartlett, 43 N. H., 224; Underwood v. People,
32 Mich., 1; Chase v. People, 40 Ill., 353; Ogletree v.
State, 28 Ala., 701; Parsons v. State, 81 id., 577; State
v. Crawford, 11 Kan., 32; Bradley v. State, 31 Ind., 485;
Wright v. People, 4 Neb., 407; Westmoreland v. State,
45 Ga., 225; Brotherton v. People, 74 N.Y., 162; Whart.
Crim. Ev., § 340; 1 Bishop Crim. Pro., § 534.

There should be clear evidence obtained on the following points, of hereditary tendencies, of sufficient predisposing causes, such as severe accident or disease, loss of
fortune, overwork, etc.; of weakness of the intellectual
powers, delusions or of epileptic fits, and previous insane acts. As to conformation of the cranium, state of
the general health and the habits of the accused, all
statements of the accused at the time of arrest, all depositions of the police and coroner, etc., should be carefully looked into by the lawyer and studied, as to the
nature of the crime, its mode of perpetration, its suddenness, the existence or otherwise of provocation and

possible motives, the time, place, etc., and the behavior of the criminal afterwards, when first arrested or in giving himself into custody.

It has been correctly said that the evidence for the insanity of a criminal — apart from those rare cases in which crime itself, or the manner of its execution, afford almost conclusive evidence of insanity — should be similar if not identical in character with that required for rendering certificates of insanity valid in non-criminal cases.   We have always taken the ground, and shall never swerve from it, never to pronounce a man or woman insane in a criminal case unless we could equally assert his or her insanity if they were brought to us as office patients, and the issues of the trial entirely indifferent to them.   We claim to have been scientific and not partisan in any and every case in which the legal profession have done us the honor to ask for our opinion and evidence.

3

CHAPTER IV.

JUDICIAL OPINIONS IN CASES OF NOTE WHERE INSANITY WAS THE DEFENSE.

NEW YORK. Oyer and Terminer, Genesee Co., Batavia, 1883. People v. Rowell. Trial for murder. Defense, insanity. "The test of responsibility is the capacity of the person to distinguish between right and wrong at the time of, and in respect to, the act complained of. He must have sufficient reason or understanding, and have an intent to do wrong and power to distinguish between right and wrong."

California. People v. Hurtado, 63 Cal., 288 (1883). Trial for murder. Defense, insanity. "If defendant was so far in possession of his mental faculties as to be capable of knowing that the act of killing was wrong, any partial defect of understanding which might cause him more readily to give way to passion than a man ordinarily reasonable cannot be considered for any purpose."

People v. Pico, 62 Cal., 50 (1882); 4 Criminal Law Mag., 281. Trial for murder. Defense, insanity. "The standard of accountability is this: Had the party sufficient mental capacity to appreciate the character and quality of his act? Did he know and understand that it was a violation of the rights of another and in itself wrong? If so, he is responsible to the law for the act thus committed, and is to be adjudged accordingly."

Indiana. McDougal v. State, 88 Ind., 24 (1883).

Trial for murder. Defense, insanity. "There can be no criminal intent where the mental condition of the accused is such that he is incapable of forming one. And hence it must appear from the evidence, beyond a reasonable doubt, that at the time of the commission of the offence charged, the mental condition of the accused was such that he was capable of forming an intent."

Karow v. Continental Insurance Co., 57 Wis., 56 ; 22 Am. Law Register, 283. The mere fact that a man commits suicide does not raise a presumption of his insanity ; but that fact, in connection with others, is pertinent to the issue of insanity.

Suicide. Coyle v. Com., 100 Penn. St., 573 (1882). Attempted suicide raises no presumption of insanity, though the fact may go to the jury with other facts.

Pennsylvania State v. Winter. Oct. 9, 1885. Northampton County Supreme Court. Assault with intent to kill. Defense, insanity. Judge Howard J. Reeder charged the jury "that with regard to the alleged insanity of the prisoner, that if they believed the act in question was the product of mental disease, they should acquit him," etc. This is humane, scientific and liberal, and we hope it will not be long before this distinguished gentleman will see his example followed in every State in the Union.

New York Oyer and Terminer, April 9 and 10, 1885. People v. Nelly Vanderhoof. Infanticide. Defense, epilepsy and insanity. Judge Van Brunt charged the jury that if they believed the act in question was due to epilepsy and mental disease, and believed in the expert opinions respecting the exculpatory effects of epilepsy, and if they believed that the disease had impaired her mind, causing mental disease, they should render a verdict of not guilty on the ground of insanity, which they promptly did.

In Alabama : In Parsons v. State, 81 Ala., 577 ; 60 Am. Rep. 193, the defendants had been convicted of the murder of Bennett Parsons by shooting him with a gun, one of the

defendants being the wife and the other the daughter of
the deceased. The defense was the plea of insanity, the
evidence tending to show that the daughter was an idiot
and the mother and wife a lunatic, subject to insane
delusions, and that the killing on her part was the offspring and product of those delusions. The trial court
followed the previous decisions of the Supreme Court of
that State (Boswell v. State, 63 Ala., 307 ; McAllister v.
State, 17 id., 434), in the latter of which the court held
that: "When the plea of insanity is interposed to protect one from the legal consequences of an act which
amounts to a crime, to render the defense available, the
evidence must be such as to convince the minds of the
jury that, at the time the act was done, the accused was
not conscious that in doing the particular act, he was committing a crime against the laws of God and his country.
If he knew right from wrong, and knew that he was violating the law, he is then guilty ; for it is this conscious
knowledge connected with the act that constitutes the
crime." On appeal to the Supreme Court, the judgment
of conviction in the Parsons case was reversed, the court
overruling the doctrine theretofore held in that State,
holding that one who, by reason of mental disease, has
lost the power of will to control his actions and choose
between right and wrong, is not responsible to the criminal law for an act which is solely the product of such
disease, although he may know right from wrong.

In California : In People v. M'Donell, 47 Cal., 134
(1873); People v. Coffman, 24 Cal., 230 (1864) ; People v.
Hoin, 62 Cal., 120 ; 16 Cent. L. J., 57 (1883) ; People v.
Hobson, 17 Cal., 424 (1861), it was held that the test of
insanity is, whether the accused, at the time of committing the act, was conscious that he was doing wrong.
In People v. Best, 39 Cal., 690 (1870), an instruction
that if the jury find the prisoner was insane at the time
of the alleged murder, they should declare him not
guilty, without regard to the degree of insanity, is
refused.

In Delaware: State v. Danby, 1 Houst. Cr. Cas., 166 (1884); State v. West, id., 371 (1873); State v. Brown, id., 539 (1878); State v. Hurley, id., 28 (1873); State v. Windsor, 5 Harr. (Del.), 512 (1851); State v. Dillahunt, 3 Harr., 551 (1840), it is held that the test of insanity is the ability to comprehend the difference between right and wrong in respect to the very act with which he stands charged.

In Georgia: The same test of insanity. Roberts v. State, 3 Ga., 310 (1847); Brinkley v. State, 58 Ga., 296 (1877); Studstill v. State, 7 Ga., 2 (1849); Loyd v. State, 45 Ga., 57 (1872); Humphreys v. State, 45 Ga., 190 (1872); Westmoreland v. State, 45 Ga., 225 (1872); Choice v. State, 31 Ga., 424 (1860).

In Kansas: The right and wrong test prevails in Kansas. State v. Mahn, 25 Kas., 182 (1861).

In Maine: The same test is adopted. State v. Lawrence, 57 Me., 574.

In Massachusetts: In the case of Abner Rogers, tried in Massachusetts before Chief Justice Shaw in 1844, he laid down the rule as follows: "A man is not to be excused from responsibility if he has capacity and reason sufficient to enable him to distinguish between right and wrong, as to the particular act he is then doing; a knowledge and consciousness that he is doing wrong and criminally will subject him to punishment; if he has a knowledge that it was wrong and criminal, and a mental power to apply that knowledge to his own case, and to know that if he does the act he will do wrong and receive punishment, such partial insanity is not sufficient to exempt him from responsibility for criminal acts." Com. v. Rogers, 7 Metc., 500 (1844); Com. v. Heath, 11 Gray, 303 (1858).

In Michigan: The same test prevails. People v. Finley, 38 Mich., 482.

In Minnesota: The same. State v. Shippey, 10 Minn., 223 (1865).

In Mississippi: The same test is followed. Bovard v. State, 30 Miss., 600 (1856); Newcomb v. State, 37 Miss.,

383 (1859) ; Cunningham v. State, 56 Miss., 269 ; 31 Am.
Rep., 360 (1879).

In Missouri: The test is the same. State v. Redemeier,
71 Mo., 173 (1879); State v. Kotovsky, 74 Mo., 247 (1881);
State v. Erb, 74 Mo., 199. Judge Henry, in the case of
the State v. Kotovsky, said that, personally, both himself
and Judge Hough did not think that the only legal test
of insanity was the ability to know the right from the
wrong of the particular act, but that one, knowing the
right from the wrong, may, in consequence of organic
mental derangement, be incapable of exercising the will,
and is therefore not amenable, criminally, for the act;
but that three of their associates were of different opinion,
and that the judgment therefore could not be reversed
for this alleged error.

In Nebraska: The same test is applied. Hawe v.
State, 11 Neb., 537 ; 38 Am. Rep., 537, (1881) ; Wright
v. People, 4 Neb., 407 (1876).

In New Jersey: The same test is applied. State v.
Spencer, 21 N. J. L., 196 (1846).

In New York: The ability to distinguish between right
and wrong, with regard to the particular act in question,
and at the very time of the commission of the act, is the
test recognized in New York. Cole's trial, 7 Abb. Pr.
(N. S.), 321 (1868) ; People v. Cavanagh, 62 How. Pr.,
187 (1881) ; People v. Devine, 1 Edm. Sel. Cas., 594 (1848) ;
People v. Griffen, id., 126 (1848); Clark's case, 1 City
Hall Rec., 176 (1876) ; Walker v. People, 88 N. Y., 86 ;
Flanagan v. People, 52 id., 467 ; 11 Am. Rep., 731 (1873);
People v. Pine, 2 Barb., 566 (1848) ; Lake v. People,
1 Park., 495 (1854); People v. Lake, 12 N. Y., 358
(1855). See, in this Lake case, the charge of the
presiding judge on the second trial. Willis v. People,
5 Park., 621 (1864) ; Willis v. People, 32 N. Y., 715
(1865); Freeman v. People, 4 Den., 9 ; People v. Waltz,
50 How. Pr., 204 (1874).   In People v. Coleman, 1 N.
Y. Crim. Rep., 1 (1881), the prisoner was indicted for
murder and the defense was insanity.   Judge Davis

charged the jury as follows: "Insanity is usually spoken of, both in common language and in the books, as a defense to crime. But it is no defense, because, where the insanity recognized by law exists, there can be no crime to defend. An insane person is incapable of crime. He is devoid, both in morals and in law, of the elements essential to the constitution of crime, and hence is an object of pity and protection, and not of punishment. * * * In the most authoritative of the English cases it is said : 'It must be clearly proved that, at the time of committing the offence, the party accused was laboring under such a defect of reason from disease of the mind as not to know the nature and quality of the act he was doing, or, if he did know it, that he did not know he was doing what was wrong;' and in a very late case, in our Court of Appeals, a charge in that exact language was held to present the law correctly to the jury," etc.

In New Hampshire: The courts, unable to find a satisfactory test of insanity, have discarded all tests. State v. Pike, 49 N. H., 399 (1870). The prisoner in this case being indicted for the murder of Brown, his counsel claimed that he was irresponsible by reason of a species of insanity called dipsomania. The court instructed the jury that " if they found that the prisoner killed Brown in a manner that would be criminal and unlawful if he was sane, their verdict should be 'not guilty by reason of insanity,' if the killing was the off-spring or product of mental disease in the defendant, neither delusion nor knowledge of right and wrong, nor design or cunning in planning and executing the killing and escaping or avoiding detection, nor ability to recognize acquaintance or to labor or transact business or manage affairs, is, as a matter of law, a test of mental disease, but that all symptoms and all tests of mental disease are purely matters of fact to be determined by the jury," etc. On appeal this instruction was confirmed. Judge Doe in this case delivered a most valuable dissenting opinion in favor of the charge of the court below on this point. "The defendant's exception," said he, "to the instructions given to the jury in relation to his responsibility as affected by dipsomania, raises the general question of the legal tests of insanity, for if the instructions given upon dipsomania are correct, they would be correct when given upon any other alleged form of insanity," etc. "A product of mental disease is not a contract, a will, or a crime, and the tests of mental disease are matters of fact," etc. "How, then, is this question of insanity to be approached by a legal tribunal? What tests are to be applied for disease? What limits assigned, within which extravagance, if thought is to be pronounced compatible with mental health," etc.? "Nor do I conceive that any tests, however elaborate, beyond the common and ordinary method of judging in such matters, would be competent to bear the strain of individual cases in the course of experience," etc. "To say that the expert testifies to the tests of mental disease as a fact, and the judge declares the test of criminal responsibility as a rule of law, is only to state the dilemma in another form. For, if the alleged act of a defendant was the act of his mental disease, it was not, in law, his act, and he is no more responsible for it than he would be if it had been the acts of his involuntary intoxication, or of any other person using the defendant's hand against his utmost resistance; if the defendant's knowledge is the test of responsibility in one of these cases, it is the test in all of them. If he does know the act is wrong, he is equally irresponsible whether his will is overcome and his hand used, by the irresistible power of his own mental disease, or by the irresistible power of another person. When disease is propelling uncontrollable power, the man is as innocent as the weapon, the mental and moral elements are as guiltless as the material. If his mental, moral and bodily strength is subjugated and pressed to an involuntary service, it is immaterial whether it is done by his disease or by another man, or a brute, or any physical force of art or nature set in operation without any fault on his part. If a man, knowing the difference between right and wrong, but deprived, by either of these agencies, of the power to choose between them, is punished, he is punished for his inability to make the choice, is punished for incapacity, and that is the very thing for which the law says he shall not be punished. He might as well be punished for an incapacity to distinguish right from wrong, as for an incapacity to resist a mental disease which forces upon him its choice of the wrong. Whether it is a possible condition in nature for a man, knowing the wrongfulness of an act, to be rendered, by mental disease, incapable of choosing not to do it and of not doing it, and whether a defendant, in a particular instance, has been thus incapacitated, are obviously questions of fact. But, whether they are questions of fact or of law, when an expert testifies that there may be such a condition, and that, upon personal examination, he thinks the defendant is or was in such a condition, that his disease has overcome or suspended, or temporarily or permanently obliterated his capacity of choosing between a known right and a known wrong, and the judge says that knowledge is the test of capacity, the judge flatly contradicts the expert. Either the expert testifies to law, or the judge testifies to fact. From this dilemma the authorities offer no escape. The whole difficulty is, that the courts have undertaken to declare that to be law which is a matter of fact," etc.

The New Hampshire doctrine is followed in Illinois and Indiana. Hopps v. People, 31 Ill., 385; Bradley v. State, 31 Ind., 492 (1869); Stevens v. State, id., 485 (1869).

In North Carolina: In State v. Haywood, Phill. (N. C.), 376 (1867), the defendant, having committed murder, and his counsel defending him on the ground of insanity, the judge instructed the jury that "if the prisoner, at the time he committed the homicide, was in a state to comprehend his relations to other persons, the nature of

the act and its criminal character, or, in other words, if
he was conscious of doing wrong at the time he committed the homicide, he is responsible. But if, on the
contrary, the prisoner was under the visitation of God,
and could not distinguish between good and evil, and
did not know what he did, he is not guilty of any
offence against the law, for guilt arises from the mind
and wicked will." On the appeal, this charge was com-
mended as a model for other trial judges in the State to
follow.

In Ohio: State v. Gardiner, Wright, 399 (1883), it
was held that the same degree of insanity which excuses
a man from his contracts will exonerate him from
accountability for crime. In the case of Loeffner v.
State, 10 Ohio St., 599, the Supreme Court laid it down
that the test was whether the prisoner had sufficient
reason and capacity to distinguish between right and
wrong, and to understand the nature of his act, and his
relation to the party injured. In the case of Farrer v.
State, 2 Ohio St., 70, the ability to distinguish between
right and wrong was the test of insanity laid down
by the Supreme Court. (See also Clark's case, 12 Ohio,
494.)

In Pennsylvania: In the case of Commonwealth v.
Farkin, 2 Pars. Eq. Cas., 439 ; 2 Clark, 208, Judge
Parsons charged the jury that "if he had reason and
understanding, so that he could judge between good and
evil, he is as much amenable to the criminal law as any
other human being," etc. Other cases where the right
and wrong test have been recognized in Pennsylvania
are Com. v. Winnemore, 1 Brewster, 356 (1867) ; Com. v.
Hart, 2 id., 547 (1868) ; Com. v. Farkin, 2 Pars. Eq. Cas.,
439 ; 2 Clark, 208 (1844) ; Com. v. Freeth, 3 Phila., 105 ;
5 Clark, 455 (1858) ; Com. v. Mosler, 4 Pa. St., 264
(1846). In the Winnemore case, epilepsy evidently
prompted the deed and the legal profession are referred
to the late Ray's very valuable remarks on epilepsy and
the legal liabilities of epileptics as contained in "Contri-

butions to Mental Pathology." (Trial of Winnemore, page 264 (1873). Because some epileptics have never committed any overt act, and may have left a brilliant record, we must not forget that epileptics have an excessive susceptibility to nervous impressions which become distorted or changed on their way to the brain, and these distorted impressions may not unfrequently prompt a criminal act. It is just as important for the legal profession to accept this scientific truth as it is for them to bear in mind that other no less important truth, that with women in the sexual evolution, in pregnancy, in childbirth, in lactation, and at the climacteric period, strange thoughts, extraordinary feelings, unseasonable appetites, and criminal impulses may haunt a mind at other times innocent and pure. Many cases of kleptomania, pyromania and of psycho-sensory insanity can thus be satisfactorily accounted for. There can be no judge who, in his heart, does not believe in a true humanity enlightened by true science. In the case of State v. Winter, tried at Easton, Pa., October 9th, 1885, the indictment being for assault with intent to kill, Judge Howard J. Reeder charged the jury that "with regard to the alleged insanity of the prisoner, that if they believed the act in question was the product of mental disease, they should acquit. That a man may have monomania and commit acts unconnected with the delusions for which he is responsible. That if he was incapacitated by mental disease from distinguishing between right and wrong in regard to the particular act, he was irresponsible," etc. Judge Reeder, in his charge, virtually takes the broad and liberal ground that where there is loss of self-control, caused by insanity, there is irresponsibility.

In Tennessee and Texas the right and wrong test is followed : Dove v. State, 3 Heisk., 348 (1872) ; Stuart v. State, 1 Baxt., 180 (1873) ; Thomas v. State, 40 Tex., 60 (1874) ; Erwin v. State, 10 Tex. App., 700 (1881) ;

Clark v. State, 8 id., 350; Williams v. State, 7 id., 163 (1879).

In the United States Courts or Federal Courts, the right and wrong test is recognized. United States v. Holmes, 1 Cliff., 98 (1858); United States v. McGlue, 1 Curt., 1 (1851); United States v. Guiteau, 10 Fed. Rep., 161 (1881).

Respecting the acts and conduct, the following has been laid down: In considering the question of the sanity of a prisoner, the jury may properly be directed to consider his appearance, conduct and language prior to the time of the commission of the alleged crime. State v. Mewherter, 46 Iowa, 88.

In a criminal prosecution for the crime of murder, the witnesses for the accused may, under the plea of insanity, be permitted to give to the jury the acts, declarations, conversations and exclamations they saw, had with, and heard the accused make at any time, shortly before, at the time of, or after the killing. State v. Hays, 22 La. Ann., 39.

Where insanity is relied on as a defense to crime, evidence of acts and conduct of the prisoner subsequent to its commission is not admissible to prove his condition at the time of the offence, unless they are so connected with evidence of a previous state of mental disorder as to strengthen the presumption of its continuance at the time of the crime, or when they indicate permanent unsoundness, which must necessarily relate back. Com. v. Pomeroy, 117 Mass., 143 (1875).

The prisoner's acts and conduct, at times other than that at which the crime was committed, are receivable in evidence. Com. v. Pomeroy, as above; State v. Kelly, 57 N. H., 549 (1876); Guiteau's case, 10 Fed. Rep., 161; State v. Hays, 22 La. Ann., 39 (1870); United States v. Holmes, 1 Cliff., 98 (1858). Where the sanity of a prisoner is at issue, a letter written by him, prior to the commission of the alleged offence, is admissible in evidence to throw light on the condition of his intellect at the time of the act

charged. If destroyed, secondary evidence of its contents may be given. State v. Kring, 64 Mo., 591 (1877);
overruling on this point State v. Kring, 1 Mo. (App.), 438,
(1876). In Choice v. State, 31 Ga., 424 (1860), it was
held that the evidence of a conversation subsequent to
the act charged, was inadmissible to prove the defendant's insanity, and so are tests made by one not an expert
at the time. In State v. West, 1 Houst. Cr. Cas., 371,
(1873), in Delaware, a valueless collection of no possible interest, which the prisoner had made from time to
time with the idea of starting a museum, were allowed,
under the plea of insanity, to be produced and shown to
the jury. In Com. v. Wilson, 1 Gray, 337, where the
defense was that the homicide charged had been committed by the prisoner under the insane delusion that the
deceased and others were engaged in a conspiracy
against him, expressions of hostile feelings toward the
prisoner made by the deceased, though not shown to
have been made in the defendant's presence, nor to
have came to his knowledge, were held admissible for the
purpose of showing the state of mind of the deceased
toward the prisoner at the time, and this tendency to
show some real ground for the prisoner's feeling toward
the deceased. In Cole's trial, 7 Abb. Pr. (N. S.), 321,
(1868), it was held that preparations made by a person to
commit the crime are relevant on the question of sanity
and premeditation.

Respecting the burden of proof, we think, as was
decided in case of Com. v. Rogers (7 Metc., 500; 1 Bennett & Heard's leading Cas. Crim. Law, 95), that a jury
may find a person insane where a fair preponderance of
the evidence is in favor of his insanity. (Also Coyle v.
Commonwealth, 100 Pa. St., 573.) In this case it was
also properly decided that where the delusion of a
person is such that he has a real and firm belief of the
existence of a fact which is wholly imaginary, and under
that insane belief he does an act which would be justifiable if such fact existed, he is not responsible for such

act. It was also held in this case, that capacity and
reason sufficient to enable one to distinguish between
right and wrong, and understand the nature, character
and consequences of his act, with mental power sufficient
to apply that knowledge to his case, furnish the legal
test of sanity. The last qualifying clause — "with
mental power sufficient," etc., etc., — is a very important
one, as many of the insane perfectly understand the difference between right and wrong, but there is the loss of
self-control produced by disease, and not mental power
enough to make them apply the knowledge of right and
wrong to their own case.

Respecting degrees of crime. In Andersen v. State
(43 Conn., 514; 21 Am. Rep., 669), it was held that
though a total want of responsibility on account of
insanity, be not shown, yet if the prisoner's mind was so
far impaired as to render him incapable of a deliberate,
premeditated murder, he should be convicted only of
murder in the second degree.

It was also held in this case that moral mania, *i. e.*,
the derangement of the moral faculties (reasoning
mania), where it is proved to exist, should be considered
by the jury in determining the degree of a crime.

It was also held in this trial that when a new trial is
asked for after conviction, on the ground of newly discovered evidence of insanity, it should be granted.

With respect to drunkenness. If a person, while sane
and responsible, makes himself intoxicated, and, while
intoxicated, commits murder by reason of insanity, which
was one of the consequences of intoxication and one of
the attendants on that state, he is responsible. (United
States v. McGlue, 1 Curt., 1), 1851. Voluntary drunkenness is no excuse for crime, but insanity produced by
continued intoxication is. Bradley v. State, 31 Ind., 492.

Where a person is insane at the time he commits a
murder, he is not punishable as a murderer, although
such insanity be remotely occasioned by undue indulgence in spirituous liquors. But it is otherwise, if he be

at the time intoxicated and his insanity be directly caused by the immediate influence of such liquors. (United States v. Drew, 5 Mason, 28.)

In the case of People v. Rogers, 18 N. Y., 9, it was held that voluntary intoxication is no excuse for crime; also, that insanity resulting from habits of intemperance, and not directly from the immediate influence of intoxicating liquors, may amount to a defense to crime.

It must be borne in mind that all alienists of experience place dipsomania under the head of periodical insanity, and do not consider it as *voluntary* intoxication at all, but involuntary. In a case of dipsomania the craving for drink which appears periodically is as irresistible to the subjects of the disease as the attack of mania is to subjects of recurrent mania, and it is just as proper to punish the one as the other for overt acts committed during the paroxysms. The public are very prone to believe that because a man or woman goes two or three months without drinking, and then suddenly plunges into a wild debauch, lasting ten days or a fortnight, that he is morally responsible; but if a physician of experience in the disease of inebriety be summoned by the court he will easily decide whether the case be one of dipsomania or of voluntary intoxication.

Respecting kleptomania. On a trial for theft, the defense being the propensity to steal, known as kleptomania, and there being evidence tending to sustain it, the court should charge the jury specifically on the point. A submission of the usual test of the prisoner's ability to distinguish between right and wrong is insufficient. Looney v. State, 10 Tex. App., 520. Judge Minklen, of the Court of Appeals, held that the trial judge should have charged the jury that kleptomania is a recognized symptom of insanity.

Respecting imbecility. An imbecile, ought not to be held responsible criminally, unless of capacity of ordinary children under fourteen years of age, *i. e.*, children of humble life and of only ordinary training. State v. Richards, 39 Conn., 591.

Respecting the effects of the opium habit on the mind, in Rogers v. State, 33 Ind., 543 (1870), the defendant sought to show the effect of such deprivation on his mental condition, but the trial judge refused to allow him. On appeal this ruling was reversed. "We think," said the Supreme Court, "the evidence was competent, as tending to show whether or not he was at the time in a condition mentally such as to be able to commit a larceny." From a somewhat extended experience in the treatment of opium cases, it is our opinion that there is a modified responsibility in many of these cases. The moral senses seem to be affected by the opium habit in many cases, and a psycho-sensory insanity induced.

Respecting insane or uncontrollable impulse: If an insane impulse leads to the commission of a crime, the actor is not responsible. An instruction that "if the jury believe that the defendant knew the difference between right and wrong in respect to the act in question, if he was conscious that such act was one which he ought not to do, he was responsible for his act, is erroneous. Stevens v. State, 31 Ind., 485 (1869).

If a person commit a homicide, knowing it to be wrong, but driven to it by an uncontrollable and irresistible impulse, arising not from natural passion, but from an insane condition of the mind, he is not criminally responsible. State v. Felter, 25 Iowa, 67 (1868).

Respecting the opinions of persons not experts. A witness not an expert may give his opinion of a person's insanity, if accompanied with the facts on which it is based. State v. Erb, 74 Mo., 199 (1887).

The opinions of an ordinary witness as to a prisoner's insanity are inadmissible. State v. Brinyea, 5 Ala., 241 (1843).

Opinions of witnesses as to the prisoner's insanity are admissible. Baldwin v. State, 12 Mo., 223.

Unprofessional witnesses may be asked, after giving

the circumstances and conduct of the party, to state their opinion as to his sanity ; and the exclusion of such evidence offered by a defendant is error. Dove v. State, 3 Heisk., 348 (Tenn., January, 1872).

The opinions of persons not experts as to the sanity of the prisoner are admissible, if accompanied by the facts upon which they are founded. Choice v. State, 31 Ga., 424 (1860).

The opinions of the family physician and of those who were in daily contact with a person who has committed an overt act, ought to be received in cases where insanity is alleged, as they are often the persons best qualified to judge of the general condition of the person for days, weeks or months prior to the commission of an overt act,

Respecting insanity in relatives. In nearly every State it is held that on the question of the prisoner's insanity it is right and proper to permit an inquiry into the mental condition of any of his immediate relatives, and this should never be lost sight of by the counsel, as it is a very important point. State v. Felter, 25 Iowa, 67 ; Bradley v. State, 31 Ind., 492 ; United States v. Guiteau, 10 Fed. Rep., 161 ; Baldwin v. State, 12 Mo., 223 ; People v. Garbutt, 17 Mich., 9 ; Hagan v. State, 5 Baxt., 615 (Tenn.).

4

CHAPTER V.

LEGAL RELATIONS OF IDIOCY AND IMBECILITY; DE-
MENTIA; DELUSIONAL INSANITY; MELANCHOLIA;
MORAL OR EMOTIONAL INSANITY (REASONING
MANIA); HOMICIDAL MANIA; SUICIDAL MANIA;
KLEPTOMANIA; PYROMANIA; EROTOMANIA; DIPSO-
MANIA; MANIA.

SECTION I.—*Idiocy and Imbecility.* Idiocy is a state
of undeveloped mind. There are no ideas or but few.
The predisposing causes of it are marriages of consan-
guinity, accidents and disease during gestation and
parturition, disease of early infant life (rickets, syphilis,
tuberculosis, struma, convulsions in dentition, etc.), and
intemperance in the parents. Idiots may possibly
recover and become intelligent. The lips of an idiot are
thick, the tongue is thick and deficient in muscular
power, the roof of the mouth is high and narrow. The
teeth come late, the lobules of the ears are either deficient or badly developed. The eyes are often affected,
there being myopia, hypermetropia and congenital cata-
ract. Stammering is very common. There is defective
size of the brain and defective gray matter, and want of
development of the convolutions. There may be either
an entire absence or a rudimentary condition of various
parts of the brain, deaf dumbness is common, and some
idiots are born deaf, dumb and blind. The higher
faculties are wanting, and the lower are very marked.
Idiots may have some degree of memory, and may be

taught to a greater or less extent. The intellectual faculties are so deficient in a medico-legal sense as to make these unhappy beings irresponsible for criminal acts which they may commit, and they are incapable of managing themselves or their affairs.

In a case of Guerniot and Broca, the brain and membrane weighed only 406 grammes, and there were atrophy amounting to 52 grammes, of the right hemisphere. Two vicious conformations which destroy symmetry, are those of scaphocephalia and plagiocephalia. The first of these conditions is a change from the natural shape of the skull, in which the parietal bones are united at the sagittal suture, so that the lateral enlargement of the brain is prevented, while that in the direction of the occipital and frontal bones is exaggerated. The parietal bones themselves are considerably increased in length. This distortion has been compared to a boat, and named from the supposed resemblance. There are forty cases on record. Virchow and Huxley hold that this deformity is the result of the obliteration of the sagittal suture, in consequence of an inflammatory action taking place during intra uterine life, which unifies and consolidates the two bones into one and prevents their lateral enlargement, and necessitates an elongation before and behind. The latter deformation—plagiocephalia—consists in an oblique or oval deformity of the cranium, in which the greatest diameter, instead of being longitudinal and antero-posterior, is oblique and diagonal ; and further, that one of the oblique diameters is greater than the other—in other terms, there is a projection of the frontal bone upon one side, and of the occipital bone upon the other. According to Topinard and Broca, the anomaly may be traced either to mechanical, pathological or posthumous causes. Such skulls as have been under observation have been generally those of the insane. These cases have been gathered from the *Annales Medico-Psychologiques* and from the bulletins of the Society Anthropologic. Bateman defines idiocy as an infirmity, consisting, anatomically, of a defective organization and want of development of the brain, resulting in an inability, more or less complete, for the exercise and manifestation of the intellectual, moral and sensitive faculties. There are various shades and degrees of this want of development from those whose mental and bodily deficiencies differ but slightly from the lowest of the so-called sound-minded, to those individuals who simply vegetate and whose deficiencies are so decided as to isolate them, as it were, from the rest of nature. The distinction between the idiot and the insane man has been stated as " *L'homme en demence est prive des brens dont il jouissait autrefois, c'est un riche devenu pauvre. L'idiot a toujours ete dans l'infortune et la misère.*" Dr. Howe, of Massachusetts, found that 99 out of 359 idiots were the children of inebriates. Dr. Kerlin gives a proportion of 38 out of 100. Dr. Beach found a proportion of 31.6 per cent. Idiots do not necessarily exhibit an obvious malformation of the cranium or skull. Dr. Bateman states that one of the most remarkable cases of idiocy he has ever seen, was that of a child with a well-formed head, remarkably handsome face, and well-proportioned body. It has been stated that 3-5 of idiots have larger heads than men of ordinary intelligence, and sometimes the brains of idiots present no deviation in form, color and density from the normal standard. Although the grey matter of the brain is connected with the intellectual power, this relation is by no means a fixed one, for richness of grey matter and abundance of nerve cells may be accompanied by idiocy; therefore, mind and matter cannot yet be regarded as identical. The mental condition of idiots is not irremediable. Under proper training, in a suitable asylum, he may become sociable, affectionate and happy. The results of education by such men as Drs. Howe, Bateman, Kerlin, Brown, Saeger, Seguin and others, have shown that by education, science has sent the idiot out into the world able to mix in society.

Imbecility is a less degree of mental deficiency than idiocy. If the imbecility dates from birth, the sensitive and intellectual faculties are somewhat developed ; sensations, ideas and memory, as well as the affections, passions, and even inclinations, exist, but only in a slight degree ; such think, feel and speak, and are capable of acquiring a certain amount of education (Bucknill and Tuke). Some imbeciles know those about them, are affectionate to their friends, but are often passionate, and are very likely to have a strong tendency to theft. They can work and can take care of themselves, but many of them are dangerous to society, as they are homicidal and apt to perform incendiary acts. There is an absence of intellectual power to a greater or less extent, conjoined with the excessive action of animal propensities. In imbecility neither the understanding nor sensibility has been sufficiently developed, while in the condition of dementia, which we shall next discuss, the patient has had these faculties and has lost them through brain disease. In imbecility, as in idiocy, the action of the mental power is disturbed by the ever-present disease, and is constantly laboring under a morbid condition, the tendency of which is to distort the moral perceptions and destroy the healthy balance of the mental faculties.

Legal Relations of Imbeciles. In determining the civil and criminal responsibilities of the imbecile, several points must be borne in mind by the lawyer. With regard to their moral sense, this class have no clear, definite idea of right, justice or law. They cannot feel for the sufferings of others. They see, only in the most imperfect manner, the consequences of their acts. They gratify every appetite or desire, regardless of consequences. Their appetites and passions are not restrained by the higher faculties of the mind, which are deprived, by disease or bad development, of their power to restrain or guide. Theft is very common with them. They have not the mental competence necessary to make them legally criminal, and it does no good to punish them in this way,

as they recommence their offences the moment they are
released from confinement, and thus are thought to be
simply wicked. Those who have strong sexual propensities soon become guilty of outrages on women, and are
imprisoned, as they are judicially decided to be rational
beings. There are many imbeciles who daily engage in
daily occupations that require no great extent of mind,
and who, perhaps, are merely thought singular by their
friends.

With respect to their civil responsibilities, if there
exists an inability of comprehending the value of numbers, the person is evidently not capable of managing
property. Ray very properly says, that the real capacity
of an imbecile's mind is to be estimated, not from any
single trait, but by a careful appreciation of all its powers,
and especially in their relation to the particular act in
question.

Relative to marriage, the person should be proved to
have had a rational idea of the marriage contract, and
of the duties and relations incident to the marriage life.
Respecting a business contract, the question would be,
had the person an adequate idea of the money involved
in the transaction? Was he independent and executive,
or was he credulous and submissive to his friends, regardless of what happened? It is no test of capacity that a
person of either sex has behaved fairly well in company,
especially when they have moved in cultured circles.
This, by constant repetition, has become automatic. Can
the alleged imbecile form a judgment respecting any new
object? How is his memory? Is he subject to gusts of
passion? Is he unfitted for all matters that require more
than a mechanical mode of action? Is he aware of his
weakness, and of the intellectual superiority of others?
Can he seize an idea so clearly as to impress it on his
mind? Is he irritable and suspicious? Has he a
clouded state of the understanding and memory? Is
he incapable of judging and deciding, when it is
necessary to weigh opposing motives? Can he express

a complex idea? Can he appreciate the circumstances that distinguish particular cases, and appreciate them according to their just value? The lawyer and jurist should carefully weigh these points when the civil responsibilities of alleged imbeciles are in question.

Respecting the *will* of a weak-minded person, if the person in question was capable of understanding its nature and effect, the instrument should be established, and *vice versa*. The question of interference or improper influence should, of course, be carefully scrutinized. (See Swinburne Wills, part 2, § 4, and 1 Story Eq. Jur., 238.) Ray says, when the mental deficiency has not been sufficient to provoke interdiction, it very properly constitutes no legal impediment to marriage; but in proof of fraud or circumlocution, the marriage has been pronounced by the courts null and void. Portsmouth v. Portsmouth, 1 Hagg. Ecc., 355; Miss Bagster's case, *ante*, § 85.

The last imbecility case in New York that greatly attracted public attention, was that of Miss Minnie A. Pancoast, who was deaf and dumb and suffered from the first degree of imbecility, whose relatives sought for a decree of nullity of marriage which she secretly contracted with her father's *masseur* Van Dorn. That this young lady was considerably below the average in point of intellect, cannot be doubted, as the evidence to that effect was remarkably strong and copious. She had very few ideas on any subject. Her intellect evidently was not strong enough to restrain or direct any tendencies of her nature. She could not reply to questions relating to any but the most commonplace subjects, even in the deaf and dumb language, and though a skilled interpreter in the sign language She was not acquainted with arithmetic, and was therefore incapable of taking care of her property. She had no judgment and reasoning power as to the marriage contract and relation. The marriage contract is a very simple one, and it does not require a very high degree of intelligence to understand it. Miss Pancoast, in our opinion, did not have either such a degree of mental capacity as to enable her to have a comprehension of the words of promise exchanged, or real appreciation of the engagement entered into; neither could she understand the nature and value of property and its management. She deserved the protection of the court and had it. The sheriff's jury and commissioners before whom the case was tried, saw at once that Miss Pancoast was incapable of comprehending the nature of the marriage ceremony and contract, and also of managing her own property; and the case was brought to a speedy termination by the graceful withdraw of the counsel of Van Dorn, upon the writer's opinion, expressed after a personal examination, that Van Dorn had no case, and that real incapacity existed, which should render such a marriage null and void. The jury rendered a verdict of unsoundness of mind.

In every such case the practical questions are: 1. Whether there are or are not such peculiarities in the conduct of the person under inquisition as are known to be characteristic of imbeciles. 2. Whether there is incompetency to manage property. 3. Whether the person, at the time of the marriage, is capable of understanding the nature of the marriage contract.

The fact of a person's being deaf and dumb certainly does not raise a presumption of insanity, and any of the deaf and dumb can legally contract marriage when it can be shown that they understand the meaning of the contract.

SECTION II.—*Dementia.* Dementia is a form of mental disorder characterized by feebleness of the ideas, affections and determinations, and by the abolition, more or less marked, of all the sensitive, intellectual and voluntary faculties (Bucknill and Tuke). It may be either partial or complete. It may gradually supervene upon melancholia, may rapidly supervene upon mania, or, by some desponding shock of the nervous system, may suddenly attack a person who has had no previous mental disease. Finally the mental faculties, the intellect, the emotions and will may become dulled, confused, and finally obliterated by reason of old age, causing senile dementia, and lastly dementia may also be associated with general paralysis, causing the disease known as dementia paralytica or general paralysis of the insane.

In dementia there is forgetfulness of the past, indifference to the present and future, and childishness of disposition. With respect to senile dementia, the case of Kinleside v. Harrison (2 Phillimore, 449), cited by Ray, in which the court decided in favor of capacity of testator, shows clearly that although a person may have lapses of memory and may show some childishness at times, yet if he can manage his affairs prudently and correctly, and in making a will show a due appreciation of the nature and amount of the property he is disposing of, and does not ignore the claims of near and dear relationship, that his testamentary capacity is to be considered good. There should be the clearest evidence of incompetency to void a will. A man's mind may be weak, and yet he may be perfectly well acquainted with the value of property, especially of his own, and may recognize his relatives and friends and be aware of the exact nature of their relations towards him and of their claims on him. The mind may be too weak to engage in large business enterprises, and yet perfectly able to understand the nature and value of property, and leave it where and to whom the testator desires. The trial judge, in cases of this kind, should carefully recapitulate, and sift, and analyze, and comment on the evidence relating to the testator's mental condition for the enlightenment and assistance of the jury. In such cases the jury attach great importance to the judge's comments on the evidence, and rightly too. The great question is, not whether a man's mind is equal to this, that or the other thing, but whether it is or was strong enough to make the will in question.

In the case of Stevens and wife v. Vancleve, 4 Wash. C. C., 262, the court said that "he may not have suffi-

cient strength of memory and vigor of intellect to make
and digest all the parts of a contract, and yet be competent to direct the distribution of his property by will."
Ray very correctly says that the great point to be determined is, not whether he was apt to forget the names of
people in whom he felt no particular interest, nor the
dates of events which concerned him little, but whether,
in conversation about his affairs his friends and relatives,
he evinced sufficient knowledge of both to be able to dispose of the former with a sound and untrammeled judgment.

Again, in the case of Harrison v. Rowan, 3 Wash. C.
C., 580, the court held : "A man may be capable of making a will and yet incapable of making a contract or to
manage his estate."

SECTION III.—*Delusional Insanity.* In delusional
insanity we have some particular delusion or false belief
impressed upon the understanding and giving rise to a
partial aberration of judgment. The person thus affected
cannot think correctly on subjects connected with the
particular delusion, and in other respects he may show
no marked disorder of the mind.

The following are some examples of the delusions that
may take possession of the mind and constitute insanity:

Mr. —— drank hard formerly. Has been temperate of
late. Did not succeed in business and fretted on this
account. Passed restless nights and became excited.
Said he was the sacred Saviour of the world. Threw
away his money, saying that it was not right for him, a
divine person, to hoard his wealth.

Mr. —— suffered much when a soldier; was over-
worked and exposed to great heat. He began to fancy
that his comrades conspired against him and attempted
to poison his food ; said that vapors of a noxious character were raised before his eyes ; said that he heard the
old woman shouting after him who had followed him

from China ; said that his bowels only acted under military orders.

In delusional insanity, or those states in which marked delusion is present, we may find melancholia, monomania or homicidal and suicidal insanity. We had, recently, a lady under our care whose life was rendered so miserable by the delusion that there was a fuse connected with concealed gunpowder in her house, and that she was at any moment likely to be blown up, that she had meditated suicide and was watching for an opportunity to successfully consummate it. This lady made a good cure in a few months, when removed from her surroundings and under proper medical treatment. Cases of delusional insanity, accompanied by homicidal impulses, are very dangerous. The case of the late Dr. L. U. Beach, of Towanda, Penn., is an example of this. Under the head of delusional insanity we must understand all the states in which marked delusions are present: Melancholia, with delusions ; monomania, with delusions ; and homicidal and suicidal insanity, with delusions.

The Trial of Dr. L. U. Beach, of Pennsylvania ; with his Psychological and Pathological History. —This case of a poor and comparatively friendless man, without the usual means of securing the favorable regards of men, calls loudly for the establishment in every State of a board of experts, who, in every trial, of rich or poor, where insanity is alleged as the defense, shall make a suitable investigation into the prisoner's mental condition. In such a case as this, such an investigation would have showed the prisoner to be a man really irresponsible, from the long-continued effect of disease, and the executive would have felt obliged to save him, in spite of the combined ignorance and prejudice of the jury which characterized the whole course of this affair.

In 1864, Dr. L. U. Beach, who was a recent graduate of the University of Pennsylvania, married Frances Sweeney, daughter of Dr. H. H. Sweeney, a leading physician of Bradford county, against the wishes of, and unknown to, her parents. Beach was the son of a well-known and highly respectable citizen of the county, but had once been under treatment of Dr. Sweeney for insanity. When the marriage of his daughter to the young doctor became known to him, Dr. Sweeney accepted the situation and took his son-in-law into partnership with him. Some years ago trouble arose between Dr. Beach and his wife, and finally she left him. He went to Hunterdon county and opened an office in Altoona. He subsequently met and married a young woman in that city, a Miss Knott by name. It does not appear that she was aware that he had a wife living; but to all appearances the couple lived very happily together. One day in April, 1884, Dr. Beach walked into the house of the young woman's brother, W. L. Knott, and coolly told him that he had murdered his wife. The brother and others hurried to the doctor's house, where they found the dead body of the young woman lying on the kitchen floor. The head was nearly severed from the shoulders. There were deep cuts on the arms, and the hands were badly cut, as though the wounds had been received while she was struggling with her assassin. A small butcher's cleaver and two sharp surgical instruments, each of the three covered with blood, lay by the dead woman's side. She had evidently first been attacked in bed, for her sleeping apartment was covered with blood, and showed signs of a desperate struggle. Bloody footprints led from her room to the room where her body was found. Dr. Beach reiterated the statement that he was the murderer, but would give no reason for committing the crime, and none has ever been discovered. The murderer was arrested and lodged in the Hollidaysburg jail. A special term of court for his trial was called in September last, 1884. Beach's defense was conducted by the Hon. Augustus Landis, on the ground of insanity.

The antecedents of the prisoner, as shown by the evi-

dence below, showed an extraordinary strong case of
mental irresponsibility and disease. In spite of the
evidence, popular prejudice so swayed the minds of the
jury that they were out but a short time and returned an
unjust verdict of murder in the first degree. A motion
was then made for a new trial, but the application was
denied.

The writer then took up the case, and, after carefully
studying the evidence adduced, made up his mind that
a great wrong had been committed. We sent an urgent
personal appeal to Governor Pattison for a commission
to be appointed to examine Beach as to his mental condition and responsibility for his acts. We set forth in
our appeal to the Governor of Pennsylvania that there
was in the Beach case a questionable verdict. That it
was a case fairly entitled to re-examination by the chief
executive of the State, as being one where popular prejudice, against the sometimes abused plea of insanity, so
influenced a jury that they rendered an unjust verdict,
and one that ought to be set aside. That the more we
had looked into the case, the more evident it was that
our statement was warranted by the facts, and that Dr.
Beach was not mentally responsible for his acts at the
time he committed the crime of murder. That because
the plea of insanity had, in the State of Pennsylvania,
been sometimes abused, was no reason for its being disregarded, and that it was better for many mistakes to
be committed than that one innocent man be wrongfully
executed. That being entirely a disinterested party,
acting entirely only in the interest of humanity and
science, our appeal ought not to be disregarded. That to
hang an insane man would be a stain on the fair fame of
the Commonwealth of Pennsylvania. Upon the evidence adduced we declared Beach, not only to be
unequivocally insane, but to have descended from a
family who were saturated with insanity. We submitted, finally, that in view of all the facts, and as we
sent sworn certificates to the Governor from physicians

who had treated Beach for insanity, that this was a case
where the hand of the law could be at least stayed until
the question of sanity or insanity could be settled, one
way or the other, by a commission of experts. We
forwarded this appeal and our proofs of Beach's insanity
to Dr. Joseph Parrish, of Burlington, N. J., the editor
of the *American Psychological Journal*, the organ of
the American Association for the Protection of the
Insane, asking his co-operation and assistance in an effort
to save Beach's life, but as we never received any answer,
do not know what, if anything, was done to co-operate
with us. The Board of Pardons, a body of laymen,
were convened by the Governor to hold a special meeting
at Harrisburg, Pa., February 13, 1885, in behalf of Dr.
Beach. After reviewing the case, this body of gentlemen, who, of course, were not competent to decide as to
the question of the existence of disease, however high
their attainments in other directions, decided not to
interfere with the decision of the trial judge, and
accordingly did not recommend Beach to executive
clemency. The Governor, in spite of our earnest protes-
tations and representations that only a medical commission were capable of determining the prisoner's mental
condition, would not further exercise his prerogative to
perform a service of the highest importance to society,
and in signing Beach's death warrant, was responsible for
his decision to the conscience and understanding of the
community ; to every man in society, who has some
regard for the triumph of right and the progress of
humanity. Beach was hanged on the 12th of February,
1885 ; a case of chronic delusional insanity from the age
of seventeen years. The evidence, as we append it,
shows his paternal grandfather to have been insane ; one
uncle who was insane ; two uncles idiotic ; a cousin who
died insane at Harrisburg, Pa., asylum ; and a maternal
uncle who died insane. Dr. Sweeney, of Clearfield, Pa.,
and Dr. Terry, of Terrytown, Pa., both sent sworn certificates of Beach's insanity.

Dr. Terry wrote us, under date of January 8th, 1885 : "Get the Governor to go to Beach himself if you can. I think if he will talk to the doctor a few minutes he can very readily see that he is not all right. If I had time I could get you more than a hundred persons that would tell you that they often thought, by the doctor's actions, that he was not just right by spells. I listened to the evidence, and I know if his trial had been in this county, he would not have had the death sentence passed upon him. Any one can see, by talking with the doctor a few minutes, that he is not all right, unless he is prejudiced."

Mrs. Frank Beach, his first wife, wrote us, "that while she had always regarded her husband as unequivocally insane, at times, ever since their marriage, that many years of intercourse proved conclusively to her that Beach was incapable of cruelty even under great provocation."

Dr. Sweeney wrote us December 10th, 1884, as follows : "Will you please inform me what is wanting in Dr. Beach's case ; one thing I positively know and assert, that he is and has been insane, and of that point I have not a doubt and can adduce positive testimony."

Dr. Beach's father wrote us as follows : "He (Beach) had three uncles that were imbeciles or insane, and his grandfather was insane on religious matters, and an uncle on his mother's side was insane in religion, and three others of his relatives further back that were not right. It seems that he had been attending a series of revival meetings, and that he joined the church the day before, and in the evening partook of the sacrament, and the next morning committed the act."

The act was the outcome of an illusion of sight. Beach's wife seemed to him an immense snake that was about to attack him, and he had an imperative conception to cut off its head, which he thought he was doing when he killed her. The following is the evidence given on the defense in the case :

The court on the Beach case was called September 3rd,

1884, all the judges on the bench. The defendant sat with the same stoical indifference that characterized him during the trial. Geo. T. Beach, a cousin of the doctor, testified of the doctor's sickness and insanity at the age of seventeen and eighteen, as indicated by his peculiarities of mind and action, also as to insanity in the family. L. L. Beach, the doctor's father, was called and testified that, in 1858, the doctor was insane ; that he had delusions ; testified that his father (the doctor's grandfather) was an insane man ; testified that he had three brothers (the doctor's uncles) : the eldest was Stephen, he was insane ; the second was Josiah, "he was regarded as without mind, he was an idiot and was so regarded by the family ; the third brother, Charles, is idiotic and so regarded ; testified that he had a sister, Ann (the doctor's aunt), who had an insane son, who died in the asylum at Harrisburg. Testified that he married Jane Grace ; she had a brother, Ambrose, who became insane ; there was another boy in that family who was idiotic or insane.

Mrs. L. U. Beach, the prisoner's wife, sworn and testified that, in 1865, she observed an attack of insanity; she said : "I might almost as well attempt to describe the showers of summer as to describe his frequent freaks of insanity." Dr. D. H. Sweeney, of Clearfield county, testified that he treated Dr. Beach for insanity. Miss Mary Sweeney, Dr. Sweeney's daughter, testified that she had seen the doctor incoherent and insane at times ; that these attacks always commenced with religious depression. Numerous other witnesses testified as to Dr. Beach's insanity and the hereditary insanity in the Beach family.

After Beach was hanged, presumably as a warning to other lunatics not to allow their delusions to control their actions, Prof. C. K. Mills, of Philadelphia, examined his brain, and our position in demanding of Gov. Pattison, of Pennsylvania, a commission of alienists fully vindicated, if, perchance, an earnest appeal for a man's life in behalf of science and humanity, needs any vindication.

Dr. Mills reported the following abnormal state of Dr. Beach's brain : "There was a lack of symmetry between the two hemispheres of the brain, an asymetry distinctly atypical. The sylvian fissure on the left side, for instance, extended upward and backwards nearly to the median edge of the hemisphere, the corresponding fissure of the right side being much shorter and about in the same position. Some of the larger fissures showed a marked tendency to confluence. Foetal and ape-like conditions were present in unusual numbers. Some of the frontal fissures and convolutions, for example, were of a low type and simple character, and one of the bridging convolutions between the occipital and parietal lobe present in the ape, but scarcely ever seen in man, was here strikingly developed. This is a rare anomaly in the human brain."

The case of Dew v. Clark, in 3 Addams, 79, gives a decision of Sir John Nicholl, pronouncing a will null and void as he considered it proved that the will was the direct product of a morbid delusion concerning the daughter. In the case of Moore, who made his will disinheriting his brothers, respecting whom he had a delusion reputed (Johnson v. Moore's Heirs, 1 Littell, 371), the court decided that Moore could not be accounted a free agent in making his will, so far as his relatives were concerned, although free as to the rest of the world, etc.

SECTION IV.—*Melancholia.* Melancholia is a state of intense depression. A state of melancholia not unfrequently precedes and ushers in an active state of mania, but the disease of melancholia is a special type of insanity from first to last. The patient affected with this disease is constantly unhappy, looks at everything through a gloomy medium, takes no enjoyment in life, and not unfrequently broods on the idea of suicide, and very often, if at home, yields to the depression, and finally attempts successfully, self-destruction. We differ entirely with those who hold that attempted suicide raises no presumption of insanity. A healthy man or woman, with a nor-

5

mally constituted mind, should be able to cope with the
reverses and griefs that life brings to all of us without
dreaming of suicide, much less attempting it. The idea
of death is repugnant to a well man or woman, and the
very fact that persons revolve the idea of suicide in their
mind is proof positive that such a person is temporarily
laboring under a disordered psychical state, a state of depression, and that the idea of suicide is engendered by
an abnormal physical state. I have more than once seen
men terribly depressed by an accumulation of oxalate of
lime in the system; have detected this by microscopical
examination of the urine ; the intense depression and
gloomy fancies and forebodings have all resulted
from a poisoning of the brain and spinal cord by
this condition of "oxaluria," and a short course of nitro-
muriatic acid in water after meals, has, by dissolving
out the oxalate of lime, completely cured my patient.
In my opinion there are very few suicides which do not
directly depend upon disordered physical states of the
system. That these physical states may have resulted
from various mental states or emotions, which tended to
depress a person, is undoubtedly true ; but faith and hope
should be brought to bear to antagonize such mental
states when resulting from loss of friends or fortune.
We see the most suicides in the hyper-sensitive and
hyper-emotional sections of society, where but a slight
incentive is needed to develop even extravagant outbursts
of excitement, and it is a rule that all purely emotional
disturbances of masses of people are attended with the
development of a higher proportional degree of insanity,
much of which ends in suicide. There is a direct dependence of the suicidal tendency on disorganizations of the
cerebral functions.

SECTION V.—*Moral or Emotional Insanity proper—
Affective Insanity or Reasoning Mania.* Every progressive student of psychiatry to-day recognizes the
existence of moral or emotional insanity proper, or rea-

soning mania (a type of insanity, in which the insane
actions and conduct are shown, rather than insane ideas,
delusions or hallucinations), as cognate with other forms
of insanity, and as exonerating patients so affected from
the accusation or punishment of guilt. Very often in
women, as a result of uterine disease, we find forms of
mental disease in which the moral and emotional part of
the brain seems to be affected. I have seen cases of this
kind frequently, in which there has been an entire change
in the character and habits of a female suffering from
uterine disorder or from puberty, pregnancy or the climacteric period. There were extraordinary acts and
conduct, false assertions and false views concerning those
nearest and dearest, without absolute delusion. We see
this also in men. It characterized the whole life history
of Guiteau, the murderer of President Garfield. The
approach of reasoning mania, or psycho-sensory insanity, which I think the best term for moral or emotional insanity proper, is said to be gradual, rather
than sudden, by those who have seen the greatest
number of cases. I know of a case whose conduct
is the result of disease, but whose acts have been
looked upon by many as signs of depravity, whose
insanity consists of false and malevolent assertions concerning her family, her physician, and any whom she
dislikes ; of plots and traps to annoy others, in which
great ingenuity and cunning have been displayed for
years ; and there is the greatest plausibility in the story
by which all such acts and all other acts will be explained
away and excused. This condition of things which came
on this patient was formerly absent, and the person is
altogether changed, which proves her insanity. She is
very acute, has no scruples about falsehood, and either
denies or justifies everything with which she is charged.
Of much of her conduct we have never been a witness,
but have received on hearsay, from credible witnesses,
much that the patient denies most strenuously. There
is a class, of whom Guiteau was an example, who suffer

from reasoning mania from their birth. They have a congenital moral defect, are from birth odd and peculiar
and incapable of acting like other people. We find
that, like the boy, Jesse Pomeroy, they are very cruel
towards their playmates, and they are generally incapable of telling the truth. They have a criminal nature,
a true moral imbecility.

The differential diagnosis which must be made between
the acts of natural depravity and the acts of the reasoning mania will depend upon the distinction between the
method, object, motive, deliberation, coolness and consistency of the acts of the naturally depraved man, and
the impulse, agitation, nervous excitement and unnatural
conduct of the reasoning maniac or case of moral or
affective insanity. This distinction has been made by
Ray, and it is a good one. In reasoning mania the
affective powers of the mind are so impaired as to over-
power any resistance made by the intellect. This is the
result of disease. During the trial of Guiteau, two gentlemen, swayed, we fear, by popular prejudice instead of
by science, were betrayed into testifying that they did
not believe in the existence of such a disease as moral
insanity. Perhaps these gentlemen would also object to
the division of the mental faculties into moral and intellectual. As against this very unscientific testimony, we
have the recorded utterances of numberless celebrated
men, from the time of Pinel and Esquirol down through
a list of names embracing Georget, Gall, Ware, Rush,
Reil, Hoffbauer, Comb, Conolly, Prichard and Winslow, down to Bucknill and Tuke, the last who have
written the best work on insanity now extant. Prichard
has perhaps given it the best definition. He defines it as
"consisting in a morbid perversion of the natural feelings, affections, inclinations, temper, habits and moral
dispositions, without any notable lesion of the intellect,
or knowing and reasoning faculties and particularly
without any maniacal hallucination." After a time the
insanity becomes well marked, and overt acts are com-

mitted which leave no difficulty in making a diagnosis. False and apparently wicked assertions concerning the nearest relatives, or plots to annoy, may constitute almost the only symptom, at times, of this form of insanity which the public can see or hear of, particularly if the nearest relatives carefully conceal from the world all outrageous conduct which is shown at home. Such patients deceive the public by their plausibility and their ready excuses for their conduct. Dr. Blandford very truly says : " When we can ascertain that the condition of things is something which has come over the patient, being formerly absent, and that a man is altogether changed, we may suspect insanity." These patients are very acute and cunning, and most unmitigated liars. There may sometimes be in the history of these cases a period, though short, of acute mania or acute melan- cholia. This may also be a precursor of a marked insan- ity, with delusions and hallucinations." In this variety of insanity a man may squander all his property or he may become a dipsomaniac. This form sometimes con- stitutes one period of circular insanity, where periods of depression alternate with those of excitement, with exaggerated conduct and absurd acts.

The responsibility of the class who have been from birth odd and peculiar, and who seem incapable of acting and behaving like other people, is sometimes difficult to estimate. They have a congenital moral defect ; they never tell the truth ; they are, so to speak, moral imbe- ciles, and it is very hard to say just how far they are responsible. They are generally the offspring of parents tainted with insanity.

My friend, C. H. Hughes, M. D., St. Louis, Missouri, says, respecting moral (affective) insanity. or psycho- sensory insanity :

" Notable instances of the subversion of mind, without accompaniment of mental perversion, are found in those cases of gangliopathy which proceed to the extent of fainting, epilepsia, chorea, etc., in which either volition or both the will and consciousness are subverted. The ganglionic (visceral) origin of certain forms of hypochondria, melancholia, and hystero-mania has been admitted since the time of Hippocrates. Morbid states of the reproductive system have long been deemed sufficient sources of certain forms of mental derangement, in which the feelings rather than the reasoning processes are disordered.

"It is conceded that kleptomania, pyromania, dipsomania, homicidal and suicidal impulses, and the morbid displays of pregnant women, and the mind disorders connected with the critical periods of woman's life, may have their starting point in uterine disorder, even with more unanimity and certainty than puerperal mania, for the latter is often as much an insanity of general hæmic and neuric exhaustion — anæmia and shock — as of reflex irritation. And, if reflex insanity be conceded, the possibility of moral insanity must be admitted, for the concession acknowledges the varying shades of mental involvement, depending upon the degree and source of the reflected irritation, from the insane longings and freaks of pregnancy to the infanticidal and other morbid impulses of *post-partum* cerebrasthenia. To concede the possibility of a homicidal or other morbid impulse not founded in delusion (and psychiatry furnishes abundant proofs of such impulses), is to admit the basis fact of moral insanity as it is clinically observable, namely, insanity not the *result* of reason perverted by disease.

"When ganglionic disease is great, and the morbid consequences profound enough to involve the intellectual faculties in marked disorder, those who deny the possibility of insanity existing without appreciable lesion of the intellect, now willingly admit the existence of mental disease, and *unwittingly*, in those minor degrees of eccentric irritation connected with the period of utero-gestation and manifested in peevishness, and insatiable longings and changes of temper, they charitably concede

that the patient is to be excused for not putting as complete a rein upon the display of eccentric feeling and
action as would be considered the proper thing in one
not *enceinte.* The intellect may appear intact or co-ex-
istent with a minor degree of moral or emotional perversion, and the perverted moral feeling excused or extenu-
ated, if indulged ; yet, if we pass a few steps further,
and venture to say that a seemingly resistless impulse,
to which the will yields while the intellect disapproves,
is insanity, then their theoretical conception of the unity
of mind — it being impossible for them to understand
how emotion, volition, and thought can be separate —
leads to the rejection of one of the most demonstrable
facts in practical psychiatry, as well as one of the most
demonstrable facts in our every-day intercourse with
minds that are not insane. Persons in the best of health
are constantly acting from impulse, prejudice, or passion,
conforming to society's usages and the dictates of fashion
or feeling without sufficient thought.

"The emotions and the intellect are not twin-born,
though they mutually influence each other. They do
not always go hand-in-hand or dwell harmoniously,
though tenanted together in the brain. In good cerebral
organizations they are often at war with each other. The
things which even sane men ought not to do they often
do, and those they ought to do they sometimes do not.

"The Apostle Paul confesses this of himself. If a saint
can concede this much of a healthy mind, a sinner can do
no less for the victim of disease. Paul was a good psychologist, and discerned, though unconscious of their
physiological foundation, the ganglionic source of certain encephalic states. He was 'constantly at war with
his members.' When he 'would do good, evil was present
with him.'

" I commend St. Paul as a psychologist to certain of our
confreres. May the convincing light of truth shine upon
them as it did upon the persecutor of the proto-martyr
on his way to Damascus, and, by way of contrition for

the wrong they have done and may yet do that least commiserated of all the mentally afflicted — the emotional, the impulsive, and the morally insane — may they speedily make amends by renouncing their heresies and, embracing the true faith, become followers of the faithful Rush, Pinel, Prichard, Maudsley, Bucknill, Tuke and Ray.

"Stephen had been stoned, it is true, but there still remained others to be saved. Many an honest Saul in our ranks, consenting to this wrong, remains to be converted. If there be any who, in perfect health, has not yielded to the dominion of impulse, emotion, or passion, let him cast the first stone at the victim of mental disease, whose intellect, while it does not restrain, yet seems not touched by the morbid process which has deranged the affections, the emotions, and the will.

" Insanity of the emotions, propensities, and passions, in which the intellect, if at all disordered, *is not appreciably so, or only momentarily so by being in abeyance or unable, through some want of connection of the will or controlling power of the latter over the impulses and passions, is a fact, however it may clash with theories of the so-called unity of mind*. It is a fact as much so as ecstasy or hypnotism, somnambulism or dreaming, which are not completely harmonious and united actions of all the mental powers. As much a fact as prejudiced, or biased, or unconscious cerebration in the healthy, working state of mind. As much a fact as the many varieties of aphasia without intellectual impairment, which the great Trousseau rejected, because he was biased in judgment by the dominant theory of Condillac and Warburton, that the mind could only think in speech. As much a fact as certain illusions or hallucinations in which the intellect does not concur, though during the formation stage of these mental spectra the reason may be in momentary abeyance. We should recognize the fact, though in so doing we may have to mend our theories or even abandon them. We should never whittle down facts to preconceived metaphysical notions.

"All observation of the varying degrees of emotional, impulsive, and intellectual life in different persons and in the same person at different ages of life attest the possibility of disorder of the emotions, propensities, or passions, without more appreciable intellectual lesion than we see in persons who are regarded as right-minded.

"Though insanity is marked generally by change of character, that change is seldom manifested in augmenting the power of the intellect and the will over the emotions or passions. On the contrary, the latter often subvert the former. Usually the disease, beginning with moral or emotional perversion, gradually involves or undermines the reason and judgment. It is thus that, in the early stages, moral, emotional, and impulsive disorder is mainly divorced from the intellect (if the two are ever then truly wedded), and what begins with an insanity of the feelings, propensities, or passions, usually goes on (if not arrested by timely medical interference) to the graver forms of more general mental involvement. These cases may even pass, if not cured, into the stages of delusions and dementia, a fact which has led some writers to doubt their existence unless associated with intellectual involvement, but which really proves the kinship of these contested varieties of mental derangement, even where neither delusion or other intellectual lesion appears, with universally recognized forms of insanity, just as the insane heredity of moral mania often establishes in our mind the fact of insanity as contradistinguished from uncomplicated vice when we are in doubt.

"Men in their sanest states are often more influenced by their feelings, prejudices, and passions than by their judgments. Insanity generally expresses itself more in action than in speech. The restlessness and constant muscular activity of many lunatics is not always the expressions of disordered intellection so much as it is an accompaniment simply of morbid feeling or irritation of psycho-motor centres, and sometimes the acts of the insane, if their after confessions in seemingly lucid intervals may be taken as even approximately true, are not infrequently independent of both conscious thought and feeling. They appear often as blind freaks of disease or mental caprice, in which the highest intellectual centres seem only unconsciously involved.

"That vigorous thinker, John Locke, who was not a mere surface observer, though he looked at insanity rather too superficially for a practical alienist, was led to the conclusion — not strictly true, but not altogether erroneous — that the insane did not so much 'appear to have lost the faculty of reasoning, but, having joined together some ideas very wrongly, they mistake them for truths, and they err as men do who argue right from wrong principles, for by the violence of their imaginations — having taken their fancies for realities — they make right deductions from them. Thus you shall find a distracted man fancying himself a king, with a right inference requiring suitable attendance, respect, and obedience; others, who have thought themselves made of glass, have used the caution necessary to preserve such brittle bodies. Hence it comes to pass that a man, who is very sober and of a right understanding in all other things, may, in one particular, be as frantic as any in Bedlam if either by any sudden, very strong impression, or long fixing his fancy upon one sort of thoughts, incoherent ideas have been cemented so powerfully as to remain united.' Locke here has reference to the deluded or delusional insane, and is only in part correct, for the insane do often both reason illogically and incoherently and establish wrong premises, from which their reasoning proceeds. Doubtless the correct reasoning manifested sometimes in the affective insanities contributed to the formation of his only partly correct opinion, for in the next sentence he says, 'There are degrees of madness as there are of folly — the disorderly

jumbling of ideas together is in some more, some less,'
and in some (he might have concluded, had he been as
familiar with the insane as they should be who aspire to
correct notions respecting them) there seems to be no
appreciable lesion of the reasoning faculties.

" Locke's idea of insanity was that it must always be
intellectual aberration, and yet his observation taught
him, despite his philosophical bias — a bias in which
many mental philosophers of the purely psychical school
still share — *that many lunatics reasoned well.* It never
occurred to him to deny the existence of insanity in
such, but to assume that they joined some ideas wrongly
together.

" With reference to another observation of Locke's, viz.,
'that reverence gives beauty and prejudice deformity to
our opinions,' it may assuredly be said with equal truth
that intellectual process in both the sane and the insane
are incited to action and influenced by moral or other
emotions, excited either by example of others or by
disease.

" Those who deny the existence of moral insanity insist
that there always exists a certain degree of intellectual
acquiescence that entitles it to be termed intellectual
insanity, though that intellectual perversion may be and
often is no greater than that which is found in the
naturally immoral and depraved : but, if one concede
this, there yet remain cases of moral and emotional disease where the intellect not only does not acquiesce in,
but actually discountenances and seeks to be restrained
from, the morbid impulse, or to subvert the morbid
feeling.

"Momentary impulses and suggestions of a morbid kind
obtrude themselves upon many healthy minds, like the
vague feelings of unreasonable unrest and depression
which obtrude unbidden into the neural chambers of the
cerebral cortex.

" Facts like these, and a hundred others needless to
enumerate, show the capability of the mental faculties

to become partially involved in aberrant action without
notable derangement of the reason.

"Men are not considered insane because they do not act
wisely ; why should it be insisted upon that the intellect
should show disorder before insanity is recognized in
those whose impulsions are undoubtedly of morbid
source, and why should the intellectual implication,
when it is found, though it be no greater than that of
some men moved by passion, be insisted upon as the
essential feature of the disease ?

"How, then, can we doubt the possibility of forms of
emotional and impulsive insanity, in which the moral
faculties are so involved by disease as to cause the individual to appear depraved ? The converse, too, is true.
There may be moral exaltation from disease as well as
from intellectual conviction, even from sexual excitation
(excessive or suppressed gratification) religious exaltation may result, as Dr. Workman and others have shown,
and that, too, without ecstatic visions or special delusions. Dr. Benjamin Rush noted long ago that a morbid
state of the sexual appetite 'becomes a disease both of
the body and mind.' This pioneer in American psychiatry and close observer of the insane, readily discerned that the will might be deranged even 'in many
instances of persons of sound understandings and some of
uncommon talents, the will becoming the involuntary
vehicle of vicious actions through the instrumentality of
the passions,' under which head he included what he
termed the lying disease, which 'differs from exculpa-
tive, fraudulent and malicious lying, in being influenced
by none of the motives of any of them.' 'Persons
thus diseased,' he says, 'cannot speak the truth upon
any subject, nor tell the same story twice in the same
way, nor describe anything as it has appeared to other
people. Their falsehoods are seldom calculated to injure
anybody but themselves, being, for the most part, of a
hyperbolical or boasting nature.' He inferred it to be
'a corporeal disease,' from it sometimes appearing in

mad people, who are remarkable for veracity in healthy states of their minds, several instances of which he saw in the Pennsylvania hospital. He recognized certain stages of intemperance as a disease of the will, and was the first to propose a hospital for inebriates, or 'sober house,' as they termed it, comparing the weakened will of a drunkard to a paralyzed limb. Rush also believed in a derangement of the principle of faith, or the believing faculty, caused by disease, also in derangement of memory, under which head he included some instances of aphasia, without the accompaniment of intellectual aberration. He was an unequivocal believer in derangement of the moral faculty, conscience, and the 'sense of deity,' and notes especially the case of a boy of thirteen years, in Bethlehem hospital, described by Haslam, 'who was perfectly sensible of his depravity, and often asked why God had not made him like other men." In the course of his life, Dr. Rush was consulted in many of 'those cases of total perversion of the moral faculties.' 'One of them was addicted to every kind of mischief. Her wickedness had no intervals while awake, except when she was kept busy in some study or difficult employment.'

"This great observer concluded that in these cases 'there is probably an original defective organization in those parts of the body which are occupied by the moral faculties of the mind,' though he could not determine where to draw the line which divides free agency from necessity, and vice from disease. He discourses further as follows :

" 'In whatever manner this question may be settled, it will readily be admitted that such persons are, in a pre-eminent degree, objects of compassion, and that it is the business of medicine to aid both religion and law in preventing and curing their moral alienation of mind.'

"Thus did one of the fathers of American medicine contribute in the beginning of the present century to the overthrow of that opprobrious doctrine of diabolical possession, or moral depravity, which has led many a hapless lunatic to the stake or the gallows, and to give us in its stead the conception of moral mania, a form of insanity just as real as the demonomania which overtook unfortunate old women past the menstrual climacteric in Cotton Mather's day, and resulted in their being drowned for witchcraft, and which, notwithstanding the reality of disease for its cause, finds even now in some quarters neither commiseration nor extenuation, being regarded as the manifestation of a wicked and devilish spirit, entitling its possessor to the punishment of the gallows or the penitentiary, rather than the restraint and treatment of the asylum for the insane. The existence of the knowledge of right and wrong with the judge, the absence of appreciable intellectual disorder with the physician, are regarded as incompatible with their ideal conception, *not of what insanity is*, but of what it to them *ought to be*, and the penalty for this theoretical misconception of the real nature of mind is visited on the unfortunate victim of disease, whose bad luck it is to be afflicted in a manner theoretically proscribed. Theoretical views and metaphysical conceptions of mind have too long stood in the way of true progress in psychological knowledge. To this has been due the fact that physical disease, as the basis of all forms of mania, now a generally accepted truth, was so long controverted. To this stumbling-block are we indebted for the inhuman treatment the insane received in the time of Galen, and up to that comparatively recent period when Pinel immortalized himself and lifted humanity to a higher pedestal by striking the shackles from the madmen in the dungeons of Bicetre.

" There is a somatic as well as a psychic element in mind as we are permitted to see it, to be taken account of in all study of psychical display, whether in health or disease, though what mind is we do not know, and perhaps we may never completely comprehend, save in its manifestations. All that we can see of mind is displayed in the operations of the intellect, the emotions, feelings and the will. There is a time in life when we see but little of the former, and a time when we see more of it than of the two latter attributes of the accompaniments of mind. The emotions and the will are part of the mind, as it manifests itself to us; and whatever may be our preconception of the impossibility of their being separated, if we see them practically severed by disease, it is only just to acknowledge the fact.

"To assert that the doctrine of moral insanity is a dangerous one, from which society may suffer, as Mayo and his followers have done, is to render science subservient to social polity, illogical, cowardly, and, of course, unscientific, whereas social polity should be ever subservient to scientific truth, whatever that may be revealed to be. Let us always speak according to our convictions. If we trim and prune truth so that we may adopt it to social expediency, we become false lights; we degrade science, the sceptre of influence falls from us, and judicial wrongs, even murder perpetrated by strong-handed law upon the weak and miserable, will continue to be committed in our name, and be the lasting monument of our disgraceful surrender of truth.

"There *is* moral perversion and degeneration resulting from disease, with but little, if any, appeciable intellectual lesion, less intellectual lesion oftentimes than we find in those whose lives have been given up to vice, through self-will or parental coercion, or evil communi cation. Then let us, when occasion demands, tell the courts so, and not say we cannot conceive it possible for moral derangement to exist without concomitant intellectual aberration, while observable facts confute such theories, and let us turn our attention to searching out, for the aid of jurists, instead of ignoring the line of demarcation between responsible and irresponsible vice; the characteristics of disease on the one hand, and on the other, voluntary moral depravity, coupled with a body sound and a mind free to choose.

"Moral insanity constitutes an observed and observable fact of psychology ; let us not seek to theorize it out of existence.

The metaphysical conception of mind, the abstraction made into an entity, as Maudsley justly observes, 'has overridden discerning observation' in some quarters, and eminent and observing men have thus suffered their judgments to become biased by the idea that the faculties of the mind cannot act separately ; that to derange one must necessarily and appreciably disorder others.

"On this reasoning, many eminent men believe the existence of moral insanity impossible, while others, among them the lamented Ray, not so biased, following in the footsteps of Prichard, who first promulgated the doctrine, see no more difficulty in recognizing insanity of the moral feelings, and of other impulses, propensities and passions, without necessary involvement of the higher faculties of reasoning in appreciable disorder, than the great Pinel did long before them in discerning what, up to his time, was regarded as equally inexplicable, namely, mania without the delirium of madness.

"Dr. Mayo, who made the first and strongest assault on the doctrine of Prichard, has unwittingly admitted, as indeed all close observers of insanity know, 'that the earliest indications of approaching insanity are moral,' and he makes the further fatal admission, 'that at every period of the actual presence' of insanity 'the powers of self-control are interfered with, the affections suppressed or altered, the passions excited or perverted.'

"All practical observers concede a frequent gradual change of feeling and conduct in prodromal insanity preceding the culmination of intellectual aberration, and some who deny the possibility of moral insanity make a classification of *moral imbecility*, concessions which logically debar all opposition to moral insanity.

"Whatever the stage at which we view mental disease, whether initial or terminal, insanity exists, as much so as fever at any stage of typhoid or typhus is fever. If a change of moral conduct have disease for its cause, it is as much entitled to be called insanity as the morbid aversions, antipathies, fears, or acts, not brought about by delusion, are to be classed among the recognized evidences of mental derangement.

"Blandford's searching analysis of Prichard's cases, while it divests many of them of the vestments of uncomplicated moral insanity, leaves a number that cannot be elsewhere placed, 'good examples of what may be called moral insanity, if the term is to be used at all,' as Blandford himself confesses, one of which he concedes deserves to be called morally insane.

"Those who engage in the study of morbid mental phenomena, with the preconception that the intellect must be always synchronously deranged in all morbid mental expression, must consistently regard every act or feeling of the insane person 'as plainly the outcome of some idea present for the moment, but present, possibly, but for the moment, and then so obliterated that the individual has lost all trace of it,' in certain morbid impulsions or feelings. Those who think 'the intellectual and emotional functions of the mind cannot be divorced, that the ideational portion of the mind is so intimately joined in operation to the emotional—the stored ideas of the brain are so influenced by the feelings of the moment, whether these arise from within or without—that the two must be sound or unsound together,' will be reluctant to concede the demonstrable fact that the affective life may be greatly changed by disease, while the intellectual processes remain intact, so far as may be discernable by any known methods of testing the integrity of the reasoning powers ; slow to recognize those cases in which the will and not the reason is weakened and perverted. Yet the morbid impulsions arising in neuropathic organisms, often reasoned against and sometimes resisted, but finally surrendered to, stand out in practical refutation of the impossible conception of the invariable unity of mind disturbed by disease.

6

"The dipsomaniacs, the kleptomaniacs, and sometimes even pyromaniacs, yielding to impulses against their reason, are examples no less destructive to this hypothesis than the auto-amnestic acts and impulses of hypnotism, somnambulism, and certain epileptoid states. To gauge insanity by the integrity or non-integrity of the reason- ing processes alone would make the automatism of cer- tain manifestations of alcoholism, epilepsia and mesmerism normal mental states. The reasoning faculties in moral insanity often appear to act as correctly as in the most perfect cerebral automatism. If there is lesion short of intellectual disease and beyond that of pure derangement of the moral faculties—and there usually is in this, as there is in all insanity, a degree of auto-amnesia by which the affected individual does not discern the change that has taken place in himself—it is simply an impared or lost appreciation of the transformation in his character, which has been brought about by disease, but many sane persons also fail to discern their descent into vicious ways. However, if this degree of involvement of the comparing faculties be deemed sufficient to ally it to insanity in general (and it does on the basis of a part of Conolly's definition), and thus to rescue a real mental disease from the theoretical assaults made upon it, we cheerfully concede it, for it is a fact that the morally insane, like most other insane persons, usually do not see themselves as others see them. But there are cases where the search must be exceedingly close to reveal any greater lesion.

"Normal mind is the sum of the aggregate display of the cerebo-psychic functions constituting the natural 'ego,' abnormal mind consists of such disorder of one or more of the cerebro-psychic functions as causes so marked a change in the psychical characteristics of the individual, whether principally involving the emotions, the reasoning powers, or the will, as to make an inconsis- tency and inharmony in the person's character explicable only by disease.

"Moral insanity is as clearly comprehended in this definition as other forms of mental derangement, and as much entitled to be recognized as a distinctive appellation and form of disease as the many other mental affections that are named on account of their prominent symptomatic feature or features.

"Not to recognize it in the present state of cerebro-mental pathology would in certain instances prove disastrous to the rights of the insane before the courts, and to their welfare elsewhere.

Legal Relations of Reasoning Mania. Ray says that general moral mania furnishes good ground for invalidating civil acts, for not withstanding the apparent integrity of the intellectual power; it is probable that their operation is influenced to a greater or less extent by a derangement of the moral power. We certainly ought not to judge the civil act by the standard of sanity, and attribute to them the same legal consequences as to those of sane men, because their real tendency is not and cannot be perceived by him. These cases, like Guiteau and others of that stamp, always think that the end justifies the means, and in my examinations of the insane I never met a single case of that kind, where fear of punishment would restrain them from criminal acts.

When a case of reasoning mania, and particularly that class of cases of reasoning mania who may properly be regarded as having a monomania for homicide, commits a crime, and the violence of the paroxysm abates, the insane man delivers himself up generally and makes no effort to escape. Sometimes, undoubtedly, the person flies from the scene and tries to escape, but most authorities on insanity unite in thinking this exceptional.

Ray says, very truly, that in homicidal insanity the criminal act for which its subject is called to an account, is the result of a strong, and perhaps sudden, impulse opposed to his natural habits, and generally preceded or followed by some derangement of the healthy actions of the brain or other organ. Taylor (643 Med. Jur.) relates the case of a young man who entered a shooting gallery, took up a pistol and deliberately shot the proprietor, who died. He said he had no knowledge of the person — he shot him simply to be hanged for it. He had been thinking of suicide for some years.

In most cases, we regret to say, moral perversion of the feelings, the result of brain disease unaccompanied with delusions, has not been held a sufficient ground to invalidate and nullify the acts of one so affected. The case of Guiteau is the last celebrated case of reasoning mania executed for murder, and we wish it might be the last spectacle of the kind on record in this country.

In the trial of Abner Rogers, for murder, Chief Justice Shaw of Massachusetts said : "If, then, it be proved, to the satisfaction of the jury, that the mind of the accused was in a diseased and unsound state, the question will be, whether the disease existed to so high degree, that, for the time being, it overwhelmed the reason, conscience and judgment; and whether the prisoner, in committing the homicide, acted from an irresistible, uncontrollable influence. If so, then the act was not the act of a voluntary agent, but the involuntary act of the body, without the concurrence of a mind directing it." (Trial of Abner Rogers, 277.)

Wharton & Stille's "Unsoundness of Mind," 43, say that in 1846 Chief Justice Gibson of Pennsylvania said in a case he was trying: "There is a moral or homicidal insanity consisting of an irresistible inclination to kill or to commit some other particular offence. There may be an unseen ligament pressing on the mind, drawing it to consequences which it sees, but cannot avoid, and placing it under a coercion, which, while its results are clearly perceived, is incapable of resistance. The doctrine which acknowledges this mania is dangerous in its relations, and can be recognized only in the clearest cases. It ought to be shown to have been habitual, or at least to have evinced itself in more than a simple instance." (Wharton & Stille's "Unsoundness of Mind,"

43.) Chief Justice Lewis, of the same State, said: "Where its existence is fully established, this species of insanity (moral) relieves from accountability to human laws." (Idem., 44.)

Judge Edmonds said in the Klein case: "It must be borne in mind that the moral as well as the intellectual faculties may be so disordered by the disease as to deprive the mind of its controlling and directing power." (Select cases, 13.) Judge Whiting, in the Freeman case (Trial of Freeman, 1847, pamphlet), expressed the same idea. In the case of Com. v. Haskell, 2 Brewster, 49), the court said: "The true test of responsibility lies in the word power. Has the defendant the power to distinguish right from wrong, and the power to adhere to the right and to avoid the wrong?"

The cases of Dean, Howson, Paparvine, Cornier and others, to be found on pp. 794-796 Woodman & Tidy's Forensic Medicine and Toxicology, are all cases of pure homicidal mania associated with reasoning.

It has been truly said that homicidal insanity has the following characters, and the lawyers and jurist can easily discover them with the aid of one skilled in psychiatry. The homicidal acts of insane persons have generally been preceded by other striking peculiarities of action, noted in the conduct of these individuals often by a total change in character.

They have often been discovered to have either contemplated suicide, or to have expressed a wish for death, or to have even wished to be executed as criminals.

Their acts are motiveless, or in opposition to the known influences of all human motives. A man known to be tenderly attached to them, murders his wife and children, a mother destroys her infant, or the victims are perfect strangers.

Their subsequent conduct is characteristic; they seldom seek escape in flight, even deliver themselves up to justice, acknowledge their crime, describe their state of mind, or remain stupefied and overcome by the horrible consciousness of the atrocious nature of their deed.

The criminal murderer has generally accomplices in vice and crime; there are assignable inducements to lead to the commission of the murder — motives of self-interest, of revenge displaying premeditated wickedness. The acts of the madman are in some instances of this character, but the premeditation is peculiar and characteristic. (Page 796, Woodman & Tidy's Forensic Med.)

SECTION VI.— *Dipsomania.* A craving after spirituous liquors is one of the recognized forms of unsoundness of mind (dipsomania), while in others it is merely a leading symptom of a mere general disorder. In some cases the craving after alcoholic liquor is intermittent, showing itself only at intervals. (See chapter VIII, remarks on dipsomania, and also chapter X.)

Delirium Tremens. As delirium tremens is a recognized disease, with mental unsoundness as one of its symptoms, the patient cannot be held responsible for his acts. Delirium is allowed to have the same effect as insanity itself on civil or criminal acts.

(See charge of Judge Gildersleeve to jury in case of McClosky, who killed her child in delirium; verdict, "Not guilty," by reason of insanity.)

Ordinary Delirium. This occurs in most severe febrile and inflammatory diseases, and is a common sequence of severe accidents and surgical operations, and it may usher in the fatal termination of chronic disorders. Delirium is almost a constant symptom of poisoning by belladonna, hyoscyamus and stramonium, a frequent result of poisoning by other narcotico-acrids; an occasional one in poisoning by the pure narcotics, and it may occur from the operation of irritant poisons.

Legal Relations of Delirium. Civil acts performed during an access of delirium are necessarily void, and criminal acts entail no responsibility. In determining the validity of wills made by persons laboring under diseases attended with delirium, the law has regard less to the proved existence of a lucid interval than to the character of the will itself. If in keeping with the testator's known character, and in harmony with intentions expressed or instructions given when sound in mind and body, if the several parts are consistent with each other, and if no improper influence was brought to bear upon him, the will should be held valid, even though the medical evidence throw doubts on his capacity. On the other hand, in the absence of these conditions, the will would generally be declared invalid, in spite of the strongest evidence of his capacity. We must distinguish delirium, with intervals of perfect consciousness, from the calmness of demeanor sometimes assumed by patients laboring under strange delusions, the result of unsoundness of mind showing itself in the first stage of convalescence from fever or other acute diseases ; or as part of delirium tremens brought on by drinking. Here the history of the case as well as the state of the patient will have to be carefully considered.

Cases illustrating the legal consequences of delirium : Cook v. Gonde and Bennett, 1 Hagg., 577 ; King & Thwaits v. Farley, ib., 502 ; Waters v. Howlett, 3 Hagg., 790 ; Bird v. Bird, 2 Hagg., 142 ; Martin v. Wolton, 1 Lee, 130 ; Bittleston, by her guardian, v. Clark, 2 Lee, 229 ; Marsh v. Tyrrel, 2 Hagg., 84 ; Hoby v. Hoby, 1 Hagg., 146.

Legal Relations of Dreaming. Very rarely the medical expert or the lawyer may be consulted concerning questions of criminal responsibility arising in those cases in which a person suddenly aroused from sleep kills another. Guy relates the case of Bernard Schedmaizig, who, suddenly waking at midnight, thought he saw a frightful phantom, which, giving no answer to his challenge twice repeated, and seeming to advance upon him, he attacked with a hatchet that lay beside him. It was found that he had murdered his wife. Ray relates

the case of two men, who, being out at night in a place
infested with robbers, engaged that one should watch
while the other slept ; but the former, falling asleep and
dreaming of being pursued, shot his friend through the
heart.   Forbes Winslow also gives the case of a peddler,
who, being rudely roused from sleep by a passer-by, ran
him through the body and killed him.   The homicidal
act in such cases is certainly not criminal, as the mental
functions of feeling and knowing, emotion and willing,
are not performed in their regular and usual manner, if
indeed they are performed at all.   In somnambulism,
also, there is either complete unconsciousness or only
such remembrance as occurs in dreaming, and when overt
acts are committed there is always a modified responsibility, and probably no responsibility at all for such acts,
whether suicidal or homicidal or kleptomaniac.   Guy
tell us of a pious clergyman who, in his fits of somnambulism, would steal and secrete whatever he could lay
his hands on, and he even plundered his own church.

SECTION VII.—*Kleptomania and Pyromania* have
been recognized by Mare, Crichton Browne, Dr. Savage,
Dr. Steinau, Tilt, Dr. Burman, Jessen and Ray, as states
of undoubted insanity.*   Bucknill and Tuke would prefer to include pyromania under the head of destructive
insanity.   That these acts may arise out of a purely
diseased mental condition there is abundant proof in the
writings of the authors referred to.   We should look in
these cases, say Drs. Bucknill and Tuke, for hereditary

---

*ₜMaré, Vol. II, p. 291, case.   Vol. II, p. 369, case.   Vol. II, p. 330,
case.   Vol. I, p. 407, case.   Ray's Med. Jurisprudence of Insanity, p.
193, case.   Klein "Annalen," XII, p. 136, case.   Maré, Vol. II, p. 309.
Bucknill and Tuke, p. 278.   Crichton Browne in Journal of Mental
Science (1860), p. 311.   Dr. Savage in " Consideration on the Cures in
Insanity," pp. 29, 30.   Dr. Hugh Miller of Glasgow, " Temporary Klep-
tomania " in " Lancet," June 15, 1878, case.   Maré, Vol. I, pp. 275, 303,
cases.   Maré, Vol. II, pp. 262, 264, etc., and Tilt's " Diseases of
Women."   Dr. Burman on "Larceny in Gen. Paralysis," Journal of
Mental Science, January, 1873.

predisposition to insanity; evidence of mental derangement prior to the development of the propensity; the
earliest symptoms of general paralysis; the occurrence
of any physical disorder, as brain fever; the suppression
of any discharge, or an injury to the head, puberty,
pregnancy; the absence (in most cases) of any inducement
to steal; the general conduct of the individual during
and after the act, and especially (although cunning and
concealment are consistent with this form of mental disorder) voluntary restitution of stolen goods. Maré says,
respecting pyromania, that incendiary acts are chiefly
manifested in young persons, in consequence of the
abnormal development of the sexual functions, corresponding with the period of life between twelve and
twenty. We should, therefore, in these cases inquire as
to whether there exist any general symptoms indicative
of irregular development, or of critical changes in the
evolution of the reproductive system, whether signs were
present before the incendiary act of approaching menstruation, its derangement or suppression, whether in epilepsy
or catalepsy, or an irregular pulse, vertigo, headache, etc.
Very often there is a change in the character, such as a
tendency to sadness, insensibility, and other symptoms
of disordered cerebral functions. Maré relates the case
of a boy who struggled for a year against such an
impulse, finally setting fire to his father's house. Ray
writes of a girl who heard voices commanding her to
burn; also of another girl who had an apparition constantly before her impelling her to pyromania.

SECTION VIII.—*Homicidal and Suicidal Mania.* Impulses leading to suicide we condone, returning a verdict
of "unsound mind," but those leading to murder we
hesitate to acknowledge as deserving any amenity in
treatment.

The majority of cases of homicidal mania, in our experience, have been among women, and as the result of grief,
anxiety, from uterine disease at the menstrual period, at

the climacteric period and after delivery, especially the last, when complicated with seduction and desertion. Women at these times are in a peculiar nervous state, not unfrequently, I am led to believe, accompanied by impulses to crime, and we do not consider them as responsible for overt acts committed at such times, especially when an overt act is antagonistic to the whole previous character of the woman. The existence of homicidal insanity ought never to be admitted without the proof of other symptoms of mental disease than the perverted instinct itself, or at least without the existence of well recognized or efficient causes of mental disease and an obvious change in temper and disposition consequent thereupon.

Suicidal Mania. This generally accompanies a condition of melancholia, and we would strongly maintain the necessary dependence of suicide on insanity. "The unhappy patients reason and struggle against the fatal propensity, but in vain. The desire to die by one's act appears to be the one mental symptom, and to present the most undoubted instance of disease affecting only one function. The majority of these cases are hereditary."

Cases Illustrating the Legal Consequences of Suicide. Suicide may not invalidate a will by raising an inference of previous derangement. (Burrows v. Burrows, 1 Hagg., 109 ; Brooks and Others v. Barret and Others, 7 Pick., 94 ; 2 Harrington, 583 ; and 2 Curties, 415.)

Chief Justice Parker of Massachusetts held that suicide committed fifteen days after the date of a person's will, was not sufficient, in the absence of other evidence, to prove him insane and thus invalidate the will.

With relation of suicide to life insurance, in the case of Borrodaile v. Hunter (5 Man. & Gr., 639), the court charged that if the deceased threw himself into the river, knowing he should destroy himself, and intending so to do, then the policy would be void ; but if he did not know right from wrong when the act was committed, then the policy would not be void. The jury found both that he intended to destroy himself and that he did not know right from wrong. Judgment was entered for the office and confirmed. In case of Cliff v. Schwabt (3 Man. & Gr., 437), the jury gave a verdict for the plaintiff, thereby deciding that a policy was not necessarily vitiated by suicide. On appeal this judgment was reversed.

Taylor Medical Jurisprudence, p. 650, 5th American edition, says truly that the term suicide in insurance policies applies, as it ought to do, only to cases in which there is no evidence of insanity. This cannot too strongly be insisted on as proper law. In the case of Breasted v. Farmers' Loan Co. (Wharton & Stille on Mental Unsoundness, p. 172), the New York Court of Appeals decided in a case of this kind, where the evidence showed that the person "was of unsound mind and wholly unconscious of the acts," that the insurers were responsible.

Although in Wisconsin and Pennsylvania the most recent judicial opinions have been that attempted suicide raises no presumption of insanity, and while such ruling is sure to be advocated by these insurance companies by whose rules the policy is made void by the act of suicide, the proper rule is that laid down by the New York Court of Appeals in the case above mentioned.

A safe rule is this : where the propensity to suicide is connected with an obviously melancholy disposition, it should be regarded as indication of mental disease. We must not lose sight of the fact, however, that while pursuing ordinary employments and avocations and manifesting very little, if any, depression, a person may have impulses to suicide which they brood over until some moral shock of domestic grief or of business reverses deprives the unhappy person of all power of resistance and the meditated suicide is perpetrated. We must remember too that Ray says that very often the patient who is prevented from committing suicide, after recovery has no recollection, or, at most, but a faint and shadowy

one of the fact itself, and believes it on the testimony of
others. I have no doubt this accords with Dr. Ray's
experience, and all his utterances I regard with the greatest respect. My own experience, however, has been
somewhat different, as cases I have attended have expressed a regret that they did not accomplish the act, and
declared their intention of accomplishing it whenever
the opportunity should offer. Two cases within my
knowledge did this very thing, being improperly guarded
by their friends. I have at present a case which has been
under my medical observation for several years. This case
would undoubtedly be taken by any insurance company
as a fairly good risk, and I doubt whether, in case suicide was committed, if a jury would not bring in a verdict of sanity. I, however, know the reverse to be true.
I know that for years, although never expressing
such ideas in society, that this person has labored
under constant depression and melancholy, has to her
husband conjured up the darkest prospects and has
constantly predicted everything of a gloomy nature.
Her melancholy mood will alternate with periods of
comparative cheerfulness, and society has never, I think,
suspected unsoundness of mind. Yet this person has
twice attempted suicide, and each time the writer was
called upon to attend the case and the person was
restored. This case I regard as a sort of monomania.
The nervous system is weak, there are cases of insanity
in the family, insomnia is very frequent, and the patient
is listless and more or less depressed most of the time.
I would not regard her as responsible for any overt act
she might commit, and have expressed fear respecting
her future. Yet it would be considered a great cruelty
to consign this case to an asylum, nor would the husband
permit it, although he appreciates fully the nature of the
case. To society at large an overt act would be the first
symptom they had noticed of mental derangement in
this case. Such a case shows clearly that we cannot
accept the verdict or opinion of any person's sanity, or

the reverse, coming from persons who only meet each other in a society way. We must search the family history and the past history from youth up of the individual and take the testimony of those who have occupied the closest relations to the accused. The testimony of an old family physician who has known all branches of the family is of immense importance sometimes, and of much greater value at time than a casual examination of one who has never seen the patient before, even though the former may know little about mental medicine.

No expert in mental and nervous diseases of any age, or experience in medico-legal trials, will either deprecate or undervalue the importance of the services which the general practitioner of medicine, or family physcian. is frequently able to render to justice in trials where insanity is alleged as a defense, or will decline to serve on a case with him. On the contrary, he will respect his opinions as those of a man who has had the closest relations with the family, who perhaps has known the person on trial from childhood, and who also knows, very likely, the whole family history and hereditary tendencies to disease in that family perfectly. The writer has frequently received from the family physician most important information, which has been of great service to him in arriving at an opinion in a perplexing case.

My experience agrees with my researches in that I firmly believe the suicidal tendency to be markedly hereditary. There is a case of this kind under our care where the suicidal tendency is most marked at each menstrual period, and we are looking forward with some anxiety to the climacteric period. There is phthisis, rheumatism and insanity in this family. There are many people who have no delusions, but who suffer from brain disease, whose only symptom seems to be uncontrollable impulse of a morbid nature. These impulses are generally recurrent with these people, but the difficulty is, that while an act of destructive impulse in a person already in an asylum is condoned, a similar act, done by a person

whose sanity has never been disputed, is visited by the
extreme penalty of the law. A great step will be gained
if the judge can be made to believe in the existence of
such a thing as uncontrollable impulse, and more importance should be attached to this matter of impulse as
regards the exhibition of leniency in trials for murder.

Respecting suicide in its relations to insanity, my late
esteemed friend, John J. Reese, M.D., Professor of Medical Jurisprudence and Toxicology in the University of
Pennsylvania, kindly contributed the following to this
chapter :

"SUICIDE   IN   ITS   RELATIONS   TO   CRIME   AND
INSANITY.—The alarming increase in the number of
deaths by suicide, within the past few years, especially
in this country, demands the thoughtful consideration
both of the philanthropist and of the legal physician.
The study of this subject presents various phrases of
great interest and importance in their medico-legal bearing; but I propose, on the present occasion, to confine
myself to the discussion of a single one — namely, 'the
relations of Suicide to Insanity.' Does or does not
the act of self-destruction always presuppose the existence of insanity ? Or, in other words, is suicide always
to be regarded as evidence of a disordered mind ? The
settlement of this question touches some of the most
vital interests of society ; and according as the verdict
rendered is yea or nay, in any particular case of suicide,
will be the feeling of the survivors that the act of the
unhappy victim was an irresponsible one, and therefore
guiltless, or that is was truly a case of voluntary, conscious self-murder, with all its appalling guilt and
tremendous consequences.

"I have no doubt that in the average mind, and
especially the unprofessional mind, there is a strong
conviction that suicide is very generally, if not always,
the result of an insane impulse ; that no person, in his
or her sober senses, could possibly commit such a terrible

act as self-destruction ; and *therefore* that the act *itself*
is conclusive evidence of insanity. This doctrine, as has
been remarked by another, ' popularly finds expression
in the verdicts of coroners' juries, who, desirous of
respecting the feelings of surviving relatives and friends,
usually return the stereotyped verdict of ' "suicide while
laboring under temporary aberration of mind." ' Indeed,
are we not told on the highest authority that ' all that
a man hath will he give for his life ? ' Can it then be
supposed that a human being, in a *rational* state of
mind, will voluntarily throw away that which is so
precious to him, and rush unbidden into the uncertainties of an unknown world ?

" It will be my purpose to show in this paper, that
while in many cases the act of suicide may be traced to a
deranged intellect, still there are instances not a few
which *cannot* be explained in this manner, but which
must be attributed simply to the deliberate purpose and
intention of the individual, the mind acting throughout
in a perfectly rational, intelligent manner, and the *will*
influenced by *motives* sufficiently cogent to impel to the
fatal act.

" I would ask attention to two distinct propositions :
(1) That many cases of suicide can undoubtedly be traced
directly to insanity ; and (2) that many other cases (and
probably the majority) *cannot* be attributed to this cause,
but are the result of an intelligent purpose and determination on the part of the victim.

" The first of these propositions is so self-evident as to
require no proofs. Everybody admits it, and indeed, as
I have observed, the popular feeling is almost universally
in its favor. Every physician of experience has had such
cases to deal with, either the result of the acute delirium
of fevers, or of chronic, morbid melancholia, in which
latter condition the propensity to suicide is so marked a
feature. It may be well to notice here that this 'suicidal
impulse,' as it is termed, sometimes displays itself very
suddenly and without any apparent previous warning,

whilst at other times it assumes the form of a delusion or
hallucination which, like an invisible spirit, may haunt
the unfortunate victim for months or years.

"A well-known striking instance of the former is related
by Sir C. Bell: 'One of the surgeons of Middlesex Hospital was in the habit of going every morning to be shaved
by a barber in the neighborhood, who was known as a
steady, industrious man.   One morning the surgeon was
conversing with the barber about an attempt at suicide
that had recently occurred, and the surgeon observed
that the man had not cut his throat in the right place.
The barber then inquired casually where the cut should
have been made.   The surgeon pointed on his own neck
to the situation of the carotid artery.   The barber in a
few minutes retired to the back of his shop, and there cut
his throat with the razor with which he had been shaving
the surgeon.   He had wounded the carotid artery in the
place indicated by the surgeon, and died before any
assistance could be rendered him.'   As is properly
remarked by the relator, 'Although this act was quite
sudden and unexpected, it may have been only the final
result of a delusion which had long existed, concealed
from others, in the mind of this man, just as the sight of
a weapon has often led to its use for the purposes of
suicide.

" I do not myself believe that this 'suicidal impulse'
ever manifests itself *suddenly and for the first time* in
a person of a *perfectly normal*, mental calibre — like a
flash of lightning out of a clear midday sky.   I cannot
but think that if the history of the individual be carefully traced, there will generally be discovered some
evidence of an antecedent, latent mental aberration,
either inherited or acquired.

" As regards those cases of suicidal impulse of a more
chronic character, such as are so frequently associated
with melancholia, I need not spend time in doing more
than merely alluding to them here.   Of course, there can

be no question as to attributing all such cases of self-destruction directly to insanity.

"Let us next examine the second proposition, namely, that suicide is very often *not* the result of insanity, but that cases frequently present themselves where the act of self-destruction was the result of a calm, deliberate determination, based on sufficient motives, and executed for a specific purpose. And here let me say a word or two about *motive* as an important factor in determining any case of alleged insanity. If we desire to form a correct estimate of a certain action or line of conduct, we usually endeavor to ascertain the *motive* which prompted thereto. The *motiveless* character of a particular act is generally regarded as strong proof of a deficiency of mental capacity; and it is constantly urged by counsel as evidence of mental unsoundness on the part of his client, whom he is trying to defend. Unquestionably, there is much truth in this assertion, as far as it goes. But before assenting to it universally, or unconditionally, we should first seek to discover whether, under the *apparent* want of motive, there may not have lain, deep down in concealment from every other human eye, and scarcely recognized by the individual himself, a motive sufficiently powerful, in that particular person, and at that particular crisis, to instigate to just such a criminal act as the one for which he is now undergoing trial.

" Prof. Casper justly remarks that we are very apt to be misled in our interpretation of criminal acts by our false notions about the *apparent want of motive* in the criminal. This point should be well noted. It is undoubtedly true that a motive to any deed may exist for *one* person, under the pressure of which he is urged on, which would be wholly inoperative in a hundred other cases. 'To recognize this, however, the inquirer must, in every case, *place himself in the position of the culprit, and divest himself of his own ideas for the time being.*" This is strong language from this distinguished medico-legal authority, but I think it is fully sustained

7

by experience. Casper cites two illustrations in support of his assertion, taken from the two extremes of social life, both of them cases of homicide, perpetrated each, under circumstances, which, had these circumstances been *reversed*, would have been attended with different results, and would *not* have terminated so tragically.

" Let us then not be too hasty in ignoring the existence of *motive* in cases of a criminal nature, whether of homicide or suicide, even though we ourselves may not be able to fathom that motive in the other, or, indeed, even to conceive of its existence. I think I am perfectly safe in asserting that we may regard the presence or absence of *motive* — of *real* motive — as the pivot, so to speak, on which will hinge the decision as to whether any one particular act of suicide was, or was not, the result of insanity. You will observe, I say the existence of a *real* motive; it is not necessary that it should be such a motive as would influence you or me, *in our present circumstances*, but a motive sufficiently constraining under the particular circumstances in which the individual was placed in that special crisis of his life.

" To take a very common illustration : A person in high social position in society, who has always enjoyed the respect and confidence of the community, has been secretly defrauding his employer of large sums of money, until concealment is no longer possible, and the terrible revelation is about to be made of the utter financial ruin of his unsuspecting victim. He forsees the near approach of the crisis which must unmask his hypocricy and expose him to the just indignation of the world, and consign him to the felon's doom. Shall he quietly wait for the fall of the avenging sword upon his devoted head, or shall he not rather save himself from the dreadful exposure, with all its awful consequences, by a swift but deliberate act of self-destruction ? Now, try this case by our touch-stone of *motive*. Will any one say that there was not motive sufficient here to account for the suicide? Certainly, there was no delusion of mind, or hallucination to interfere with the man's reasoning processes, or to pervert his volition. The motive was based on a *reality* — on a real estimate of his position and its consequences. It was purely a question of *choice* between two great evils — loss of honor or loss of life ; and he deliberately chose what *he* considered the lesser, and so died by his own hand. Certainly, we cannot ascribe such a suicide to insanity. Does any one doubt that the anarchists of Chicago would not gladly have forestalled their merited doom, if it had been possible, by some dynamite or other mode of self-destruction, as indeed did happen to one of their number ? In this same category we may class those cases of suicide following loss of property, loss of employment, loss of honor, mortified pride, disappointed love, disappointed ambition, and others of a similar nature. Take away from these wretched ones all idea of a future responsibility and belief, and I ask what is there to hinder them from putting a speedy voluntary end to all their sorrows and disappointments ?

" I know very well that some will urge that all such unhappy persons have been driven into temporary insanity by their various sorrows and afflictions, and that the suicidal act was committed while under the insane delusion. The reply to this apparently plausible, though really specious objection, is, that it is 'begging the question' to *assume* the very thing that is to be proved. Such a line of argument would equally excuse any and every act of violence inflicted upon another, upon a similar plea of insanity.

" Perhaps it might be well to fortify this assertion, that suicide is not necessarily connected with insanity, by an appeal to history, past and present, sacred and profane. It is well known that in the palmiest days of Greek and Roman civilization, the universal sentiment of the people, fostered by the Stoic and Epicurean philosophy, was in favor, not only of the *lawfulness* of suicide, but of its *propriety* and *excellence*. The philosophers taught that a voluntary death was always to be preferred to disgrace,

or even to personal suffering. Thus Leno, the founder
of the Stoics, hanged himself in his 98th year, rather
than endure the pain of inconvience of a dislocated joint.
The generals, Hannibal and Mithridates, Mark Anthony,
Brutus and Cassius, sought a suicidal death after military defeat. Demosthenes, Themistocles, Cato, of Utica
and Seneca, as also Cleopatra, believed in suicide and
practiced it.

"Sacred history affords us examples of self-destruction
in the persons of King Saul of Israel, of the mighty
Sampson, of Ahithophel, the counsellor of David, and of
the traitor Judas. No one for a moment would think
of ascribing any of the above instances of suicide to
insanity. Certainly those concerned, comprising philosophers, statesmen, generals and commoners, were
persons of no mean order of intellect.

"In more modern times, there have not been wanting
persons of brilliant intellect, though of skeptical belief,
who have openly advocated both the lawfulness and the
desirableness of suicide, among whom I may mention
Rousseau, Madame de Staël, Hume, Gibbon, Montaigne
and Montesquieu. However we may condemn the doctrines of these people, we cannot surely ascribe their
false notions to any want of intellect, since they shone
as stars of the first magnitude in the literary and social
circles of the day.

"Even in our own times, in some of the oriental
countries, the practice of self-immolation still prevails to
a certain extent. The custom of the Hindoo widow
burning herself upon the funeral pyre of her husband, I
believe, still lingers in certain parts of India; and the
singular custom of *Hari-Kari* is scarcely yet entirely
abolished in Japan.

"The benignant influence of the Christian religion, as it
gradually spread over the world, had the effect of entirely
changing the popular sentiment in regard to the practice
of suicide. The old heathenish notion of its *lawfulness*
was abandoned, and it came to be looked upon as a crime

of the first magnitude and to be punished as such by the laws of the country. Indeed, so severe were the penalties visited upon the person, the effects, and even the descendants of the *felo-de-se*, that it led to a recoil of popular feeling against the barbarous and revolting laws of the age; and the consequence was an attempt to ascribe these cases of suicide to a *disordered mind*, so as to avoid the implication of crime, and thus to remove them from all responsibility. Now, you will observe that this very fact of suicide being regarded by the laws both of the State and the church as *criminal*, and to be punished accordingly, proves conclusively that these laws could not and did not regard it as originating in insanity, since insanity necessarily precludes all idea of responsibility; and therefore all such laws would be most monstrous and inhuman on such a supposition. In England, it is only since a little over fifty years that the bodies of suicides were allowed burial in the parish church yards, and up to about that same period the practice still prevailed in that country, of burying a suicide at the cross-roads in the country, and thrusting a stake through the body, in order to show the public detestation of the crime.

"At the present day this question of 'the relation of suicide to insanity' occurs chiefly in cases of life insurance, where the payment of the policy is contested by the company on the ground of the *voluntary*, intentional death of the assured *while in his sound mind*. Until within a few years past it was the practice of all the life insurance companies to insert in their policies a saving clause exonerating them from payment, *if the death of the assured was caused by his own hand, whether sane or insane*. This was manifestly both unjust and unlawful, so far as the *insane* act of self-destruction is concerned; for evidently a person should be held no more responsible for the loss of his life, whether he takes it himself when impelled by the hallucinations of melancholia, or under the ravings of the delirium of typhoid fever, or of inflammation of the brain, than if he dies from an ordinary disease or from an accident. If, on the other hand, it can be clearly shown that the assured, after placing enormous amounts of insurance money upon his life in various offices, and after making only one or two payments of premiums, had committed suicide in a perfectly *intelligent* manner, and evidently with the intent and design of benefitting others pecuniarily by his death, *then* the question assumes altogether a different aspect. Neither justice nor equity could plead for favor in such a case. We certainly think that in such a case, unless it can be clearly demonstrated that the individual was insane, the company should *not* be forced to make payment. Let us try such a case by our touch-stone of *motive.* Why should not such an individual, pressed down by the burden of debt, which he has been vainly striving to pay, if restrained by no belief in, or dread of, a future retribution, why should he not embrace the tempting offer to insure his life for an amount that will not only repay his indebtedness and rescue his name from dishonor, but at the same time save from poverty and want those whom he loves better than himself? Why should not the spirit of self-sacrifice be as dominant in such an one as in any of the instances which I have adduced from ancient times? The true issue in all such cases is : ' Did the assured intend, freely and intelligently, to destroy himself?—*intelligently,*— *i. e.*, with his mind at the time unswayed by any insane delusions?' I pray you to observe the very important distinction between doing a thing *intentionally*, and doing it *intelligently.* The insane man, equally with the sane one, commits the act of self-destruction knowingly — *intentionally* — with the full purpose in view of terminating his existence ; but he does not do it *intelligently*—with his mind in its normal equipoise — ' unswayed by an insane delusion.'"

Legal Relations of Homicidal Mania. The legal test of insanity in criminal cases, in cases of homicide, should be, the existence of any subjective morbid condition of the nervous system which misleads the mind or conduct. The basis of insanity consists in the changing and misleading subjective impressions of the insane person, coupled with the resultant change of conduct or of reasoning, or both. There is a change of mental character, as compared with former self or normal ancestral type. I fully agree with Dr. Hughes that physical disease, sickness, impresses itself on the conduct or character of the person affected by it, misleading and perverting him in the exercise of his psychic powers. We must recognize, as a sick man, one who has the undefined perversions of feeling displayed in melancholia and homicidal and suicidal impulses, and also the kleptomaniacal, pyromanical, nymphomaniacal, and other erratic feelings which mislead the judgment and conduct of the insane.

Violent homicidal impulses are very common in the epileptic, sometimes preceding, sometimes following the fits, and sometimes taking their place (masked epilepsy).

Imbeciles are peculiarly liable to impulses to murder and often give way to the uncontrollable impulse.

There is an instinctive or impulsive mania, and Guy correctly states that the homicidal acts committed under its influence have most or all of the following characters: They are without discernable motive, or in opposition to all known motives. A man kills his wife, to whom he is tenderly attached, a brother his sister, a mother her infant. The victim may be somebody whom he never saw before, and against whom it is impossible that he should have malice. The victim of this blind passion may be an animal incapable of offense. After the commission of the act, he does not seek to escape; he often tells what he has done. He does not conceal the body, but openly exposes it. He delivers himself up to justice. He tells of the state of mind which led to the act, and either remains stupid and indifferent or is overwhelmed by remorse. He has no accomplices, has made no preparations, and takes nothing from his victim. Perhaps he has told of his strong impulse to kill, and has begged to be restrained. These homicidal acts are generally preceded by a striking change of conduct and character, and on inquiry the accused is often found to have an hereditary tendency to insanity, to be subject to fits, to have attempted suicide, or to have wished for death.

Homicidal Mania Resulting from Religious Insanity. These cases are constantly under excitement or depression, and are subject to illusions and delusions. "They transform the persons with whom they are associated into supernatural beings, endowed with authority or power not to be questioned or resisted; and they commit common and familiar sounds into the articulate language of temptation or command. One religious maniac therefore kills a relative or keeper, imagining him to be a fiend; and then thinks that he has a direct commission from the Deity to fulfil some mission of wrath or extirpation. In case of religious mania, we can never safely affirm that the homicidal act was not the natural consequence of a command which the maniac would deem it impious to resist, or of a delusion which places him, in his own sincere conviction, beyond and above the operation of human laws. The maniac who believes himself to be God or Christ would, from the very nature of the case, deem himself irresponsible."

Homicidal Mania from Jealousy. In these cases the jealousy has shaped itself into a distinct delusion, and Guy correctly states that they are such acts as, if committed by sane men on the evidence of their senses, would be punished as manslaughter and not as murder.

Homicidal Mania from Domestic Anxiety Exaggerated into Fear of Starvation. Those parents who kill their children are generally noted for their domestic virtues and great attachment to their victims, and there is not one point of resemblance between those insane murderers and ordinary criminals.

Homicidal Insanity from Delusion. A homicide, the result of a delusion on the part of the insane man, may be accomplished much as a sane criminal would do it. Deliberation, forethought and preparation may all enter into the accomplishment of the deed, but we must not infer that therefore the insane man had such an amount of self-control as will prevent the homicidal deed. He has not. We must not confound the act itself with the mode of accomplishing the act.

To leave the subject of homicidal mania arising from morbid motives and delusions, and return to it as a monomania or as simply an irresistible desire to kill, it is of importance to bear in mind that it may co-exist with general defect or disorder of mind. It is of the greatest medico-legal importance to know that there may be outbursts of maniacal fury with homicidal impulse and no reliable proof of any prior history of mental disease.

The most distinguished physicians, devoting their time exclusively to treating mental and nervous diseases, both in Europe and in America, concur and unite in their belief that the insane and irresistible impulse prompting to murder and destruction, which has been designated homicidal monomania, is a distinct disease, from which even childhood is not exempt. The powerful impulse to kill may originate in morbid motives — in delusion — or a mere blind, motiveless impulse to kill is felt against which the monomaniac himself strives most earnestly. Bucknill and Tuke, Crichton Browne, Dr. Skae, Wilks, Woodman and Tidy, Guy, Krafft — Ebing, Castelnau, Devergie, and every physician of note as an alienist, all agree as to the existence of homicidal mania as a distinct physical disorder, and, as it is a fact of science, law should also recognize it.

SECTION IX. — *Legal Relations of Mania.* It is important for the lawyer to bear in mind the following points relating to mania :

1. In mania, consciousness, memory and reason, may remain intact even in the midst of the most violent paroxysms. The patient is whirled about in an emotional storm.

2. The maniac's senses are deceived and confounded. Illusions and hallucinations of sight are very common.

3. The persons with whom the insane man associates are apt to derive their characters from his delusion.

4. Real impressions on the organs of sense become, as in dreams, the materials of imaginary scenes.

5. The strange antics of the insane man are the effects of his delusion.

6. The violence of the madman is often not the effect of mere passion, but of his delusion.

7. The maniac, if of a reserved disposition, or when impelled by a strong motive, can conceal his delusion.

8. The acts of the maniac often evince the same forethought and preparation as those of the sane.

9. The maniac, in spite of his cunning, is easily imposed upon and managed.

10. Maniacs in confinement are often conscious of their state, and know the legal relations in which it places them.

In deciding medico-legal questions, it is quite necessary to know that these are some of the leading characters of mania.

The lawyer should be aware of the fact, which at times is of considerable medico-legal significance, that mental disease may arise as a result of the revolution of the system in either sex, occurring at puberty. Masturbation may also cause it; uterine or ovarian disease likewise. The condition of gestation or pregnancy may give rise to it. Mental disease may appear after parturition, when it is termed puerperal insanity, and it is at all uncommon in a mild form. Obscene words and self accusations of impropriety and delusions connected with sexual matters are all common at this time. The period of lactation may be associated with mental disturbance. The climacteric period in women not uncommonly gives rise to nervous troubles which may end in insanity. Insanity may be among the sequellæ of fevers. The rheumatic and gouty poison may cause insanity at times. Syphilis and phthisis may both give rise to and be associated with mental disease, and alchoholic insanity is very frequent, and is accompanied by hallucination of hearing and other hallucinations. Finally, we have the insanity of old age or senile dementia.

Mania Transitoria or Temporary Insanity. In these cases, which are of great interest, as they are often before the courts for the commission of overt acts, hereditary predisposition is to be placed at the head of causes, for, as Marc says of insanity in general, it plays so marked a character in the production of this malady, that whenever there is a possibility, in a medico-legal investigation, of demonstrating its existence, it is sufficient almost of itself to establish the reality of a lesion of the understanding, or to weaken considerably the possibility of its being feigned.

Dr. A. Devergie, a most eminent alienist of France, in a paper read before the Imperial Academy of Medicine, entitled, "Transitory Homicidal Mania: When does Reason End or Mania Begin?" in the "Journal of Psychological Medicine and Mental Pathology," No. XVI, says: "Those physicians who have devoted themselves to the treatment of insanity, admits that besides dementia, mania and monomania, there exists an instantaneous transient insanity, which they call transitory, and as the result of which, an individual, until then, in appearance at least, of sound mind, commits suddenly a homicidal act, *and returns as suddenly to a state of reason.* It would be easy to quote a hundred authors of recognized pre-eminence in psychological medicine to the effect that such an affection as temporary insanity really exists. The authorities on medical jurisprudence are likewise decided upon this point, and the fact is accepted every day by courts of law. It is unnecessary, therefore, to adduce further support to the doctrine."

Dr. Wm. A. Hammond says: "There is a form of insanity, which, in its culminating act, is extremely temporary in its character, and which, in all its manifestations, from beginning to end, is of short duration. This species of mental aberration is well known to all physicians and medical jurists who have studied the subject of insanity. By authors it has been variously designated as mania transitoria, ephemeral mania, temporary insanity and morbid impulse. It may be exhibited in the perceptional, intellectual, emotional or volitional form, or as general mania. The exciting causes of temporary insanity are numerous. It may be induced by bad hygienic influences, such as improper food, exposure to intense heat, cold or dampness, or to a noxious atmosphere ; by undue physical exercise, by disease of the heart, by blows upon the head or other parts of the body, by certain general and local diseases, by the abuse of alcoholic liquors, by the ingestion of certain drugs, such as opium, belladonna and hasheesh, by excessive intellectual occupation, by loss of sleep, and, *above all, by great emotional disturbances.**   Among these latter, religious excitement, grief, disappointed affection, *and especially anxiety, by which the mind is kept continually on the stretch, tortured by apprehensions, doubts and uncertainties, by which it is worn away more surely than by the most terrible realities.** The predisposing causes are to be found in the individual as an inherent part of his organization. They consist in a hereditary tendency to insanity, or to some other profound affection of the nervous system, or of an excitable nervous temperament, which is incapable of resisting those morbid influences which persons of phlegmatic disposition would easily withstand. Thus all men are not affected alike by disturbing causes, because all men are

---

* Italics are the authors'.

not cast in the same physical or mental mould. A circumstance which will produce insanity in one person will scarcely ruffle the equanimity of another. The immediate cause of temporary insanity is the disease itself, of which the mental aberration is simply the manifestation. No fact in medical science is more clearly established than this, of the action of the emotion on the circulation of the blood in the brain. This form of insanity is known as transitory mania.

"It may be defined as a form of insanity in which the individual, with or without the exhibition of *previous notable symptoms*, and with or without obvious exciting cause, suddenly loses the control of his will, during which period of non-control, he commonly perpetrates a criminal act, and then as suddenly recovers, more or less completely, his power of volition. Attentive examination will always reveal the existence of symptoms precursory to the outbreak which constitues the culminating act, though they may be so slight as to escape superficial examination."

Dr. Jarvis, in a paper published in the "American Journal of Insanity," for July, 1869, says of mania transitoria : "This is a form of mental disorder which suddenly appears in a person previously sane or not supposed to be unsound in mind : it has a short duration and suddenly disappears. This is not exclusively a new or an old doctrine, but it has been taught in France and Germany and other countries and by managers of the insane and by writers on these topics. It is recognized by the psychological authorities of Great Britain, and is admitted by courts and juries having the management of persons who have committed acts which would otherwise have been considered as criminal, and for which they would otherwise have been doomed to death by the scaffold."

Dr. Castelnau, in the "American Journal of Insanity," concludes that "there exist instantaneous changes in the mental faculties." "Mania instantaneous, temporary, transitory, fleeting, a mental disorder which breaks out suddenly like the sudden loss of sense by some physical disease; the subject is urged in a moment to automatic acts which could not have been foreseen."

The late Dr. Ray, an alienist of the very highest character, said: "Yet sometimes, especially on the operation of a powerfully exciting cause, it breaks out suddenly and terminates in a few hours. It has been called transitory mania, or instantaneous mania." Again he says, in cases like that of Mercer: "When a man destroys the seducer of his wife, sister or daughter, we often see the influence of the insane temperament, and the effect has been very much in determining the quality of the act. We also know, as a matter of no very infrequent experience, that insanity may be produced instantaneously by a profound moral shock. If a person might be deprived of his senses on a piece of good news, or of the death of one very near and dear, is it strange such results would follow what is calculated, above all others, to stir the soul to its inmost depths? What the mental condition actually is must be determined by evidence in the case, and any doubt there may be, we may be quite sure, will be given in favor of the accused."

All writers on medical jurisprudence and insanity, whose opinion is worth quoting, concur in the existence of mania transitoria, and personally I have seen so many examples of it that I should as soon be incredulous about the existence of typhoid fever as of the existence of transitory mania. Those persons are the most apt to exhibit mania transitoria who exhibit a predisposition to insanity, whose general health is impaired from any cause, who are naturally nervous and excitable, who have been subjected to any great trial of their feelings in any way to cause them to dwell much on the subject, to lie awake nights on account of it, and then have added the application of a great and sudden excitement or some strong emotion; as great emotional disturbances, are the most powerful exciting causes of this form of mania. I think a careful alienist could, in these cases, if he could see them before the commission of an overt act, always detect premonitory symptoms of an attack, but it is not at this stage that a person receives any medical thought or supervision. The act of violence is the first manifestation of the disease of the body affecting the mind by deranging its functions, which constitutes insanity, that the public sees, and prejudice or ignorance very often denies, what to a physician is perfectly evident.

In transitory mania, the control which the intellect normally exercises on the will is, for the time, destroyed, and the overt act is the result of an automatic impulse. It appears without premonition, and disappears as quickly as it came. It is a discharge, in the convolutions of the brain, perfectly analagous to that of epilepsy, and very frequently there will be no more recollection of the occurrence than the epileptic has. Indeed, a case of "petit mal" is the very case to be transformed on the moment into a case of transitory mania upon the application of even a slight exciting cause, and the existence of the lesser form of epilepsy should be diligently sought for by both physician and jurist in every such case. Many women, while menstruating, have their nervous systems so overwrought that any great emotional disturbance would be very likely indeed to precipitate an attack of instantaneous mania, especially if such a woman had inherited a predisposition to some form of nervous or mental disease, and there would be perfect irresponsibility for any overt act committed during such a state, in our opinion. Finally, I do not myself believe that transitory mania ever manifests itself suddenly and for the first time in a person of a perfectly normal mental calibre—like a flash of lightning out of a clear sky. I cannot but think that if the history of the individual be carefully traced, there will generally be discovered some evidences of antecedent latent mental aberration, either inherited or acquired.

As to What Constitutes Sanity.   Judge Emunds once gave this very able definition of sanity : "A sane man is one whose senses bear truthful evidence ; whose understanding is capable of receiving that evidence ; whose reason can draw proper conclusions from the evidence thus received ; whose will can guide the thought thus obtained ; whose moral sense can tell the right and wrong of any act growing out of that thought ; and whose acts can, at his own pleasure, be in conformity with the action of all these qualities.   All these things unite to make sanity.   The absence of them is insanity."

Recorder Hackett, in the case of The People v. McFarland, 8 Abb., N. S., 92, said : "A state of sanity is one in which a man knows the act he is committing to be unlawful and morally wrong, *and has reason sufficient to apply such knowledge and be controlled by it.*"

Most maniacs have a firm conviction that all they feel and think is true, just and reasonable, and nothing can shake their conviction.   .

CHAPTER VI.

AID TO THE LAWYER IN ARRIVING AT A JUDGMENT AS TO SANITY OR INSANITY IN CASES WHERE INSANITY IS ALLEGED AS A DEFENSE.

THERE is probably no disease which presents greater difficulties in the way of diagnosis, than insanity. In most diseases we examine physical signs and symptoms, and we determine by our senses the existence of such diseases. In insanity, on the contrary, we have to be guided chiefly by our knowledge of the normal functions of the mind, and in our examination we have to rely on our intellect rather than on our senses, although of course the latter are called in to assist us. It is, however, very often extremely difficult to decide with certainty, as we are expected to do, as to the existence of mental disease, and we assume a great responsibility whichever way our decision may be given. We either give the patient liberty to take his place in society, and thus expose society to the consequences if he prove to be insane, or we place him in confinement in some institution for the treatment of the insane, thus depriving him of his liberty and his family of his support.

It becomes, then, a matter of great importance to decide rightly as to the existence of mental disease, for if this is not rightly done, we shall expose ourselves to the risk of great mortification, and also to the loss of professional reputation. Before going to see a patient who is to be examined for the existence of insanity, it is advisable to find out all one can from the friends and relatives; but

8

in accepting such statements it is wise to allow a wide margin for their information in regard to hereditary pre-disposition, as most people, foolishly considering the existence of insanity in their family a disgrace, will pertinaciously conceal and deny this fact. Another reason for this concealment may be, that the members of such families are not infrequently odd and eccentric in their behavior, even when perfectly sane, and do not care to have their peculiarties attributed to hereditary taint of insanity, and therefore endeavor to mislead their physician on a point which is to him of the utmost diag-nostic importance. Indeed, this and the question of previous attacks, are perhaps the two most important points in the diagnosis of any given case.* We should endeavor, when we are called to our patient, to gain his confidence, and, from a general conversation, lead him cautiously to his state of health and mental feeling. If we are abrupt and wanting in tact, we shall probably defeat our object, and the patient, if displeased, will either refuse to listen to or answer our questions, or will become very angry at our conspiring to deprive him of his liberty. If we are fortunate enough to get a his-tory of the patient, we can generally determine easily the existence or non-existence of insanity by the patient's

---

* *Premonitory Symptoms of Insanity.* Before a previously healthy person becomes insane, we shall generally find that he has manifested depression, unwonted excitability, disregard of the minor proprieties of life, a change coming over the warmest affections, quick changes and rapid transitions in the current of the feelings, sleeplessness, and a complete change of the character and habits, the person, meanwhile, entertaining no delusions, but occasionally losing his self-control, the general air and manner at such times being strongly expressive of the inward emotions; intervals of perfect calmness and self-control, dur-ing which the person clearly discerns his true relations to others, and even, perhaps, recognizes the influence which the incipient disease exercises over his feelings and actions, with finally, the utter down-fall of the intellect, manifested by the fury of mania or the moodiness, suspicion, depression and impulses toward self-destruction of melan-cholia. All these are the successive links forged in the chain of insanity.

appearance and conversation. Many times, however, we have to rely alone on the conversation, general appearance and conduct of the patient, unaided by any other resources. After having gained our patient's confidence, and having drawn him into a pleasant conversation, we should first inquire about previous attacks, then into his hereditary history, then into any predisposing causes, such as intemperance, vocation, habits, etc., which may have operated in the production of insanity. Also as to injuries to the head or spine which may have occurred, sunstroke, etc. We should then systematically, but carefully and cautiously, examine into the vegetative and reproductive functions, and then carefully examine the nervous system for the existence of such lesions as paralysis, epilepsy, catalepsy, hysteria and allied affections. We should next examine the different senses, beginning with sight, and in this way we shall find out if our patient has good vision, if the retina is normal, and, what is more important, we may discover if he has hallucinations or illusions pertaining to this sense. We may then proceed to the sense of hearing, examining for deafness, and also to discover any hallucinations or illusions of hearing. Proceeding to the sense of smell, we shall discover if it is normal, and also if there are any hallucinations or illusions connected with it. Taking up the sense of taste, we may inquire as to the existence of hallucinations or illusions. Patients often complain of their food being poisoned, or that they are eating injurious and hurtful things with their food. The last of the senses, that of touch and nervous sensibility, may be examined for imaginary sense of pain, the existence of reflex action, hyperæsthesia, and lastly, for hallucinations and illusions pertaining to this sense or referring to internal organs of the body. The mental symptoms unconnected with the special senses and pertaining to the intellect, the emotions, or the will, may finish the examination. Whether the diagnosis of insanity present itself to the physician in a purely medical or in a medico-legal point of view, the principles of diagnosis are the same, and we must pursue our examination in precisely the same manner. The first thing we are generally called upon to decide is, whether the patient can be treated at home, or whether it is necessary to place him in an asylum, and we are also probably asked for a prognosis, which latter cannot be too guarded, whatever may be our own impression at the time, about the patient. Let us consider for a moment the first question, that of the propriety of removing our patient from his own home, either to some private retreat or to a public asylum. For those who can afford the expense, I prefer a residence away from home in some private retreat, where but few patients are admitted, for the reason that they unquestionably can have much greater care and attention bestowed upon them than in the congregate plan of treatment. If they cannot afford this, a residence in any well regulated public asylum, where, as a rule, the superintendents are earnest, thoughtful men, careful for their patients' welfare, is to be desired *as soon as possible*, while the disease is in its early curable stages.* Insane patients are, by the very nature of the disease, inclined to do mischief. They are controlled in their actions by delusions which are to them vivid realities, and no one knows what they may consider it right and proper to do when under the influence of such delusions. Some of the most fearful crimes have been committed by those who have previously been regarded as harmless patients, and no one, therefore, should take upon himself the responsibility of advising that a patient whom he is called to see should be kept at home. The mere

---

* It should be borne in mind that much of the popular prejudice against hospitals for the insane springs from unfounded statements made by persons who have been inmates of such institutions, and who have been discharged before they were fully restored to reason. A person who has made a complete recovery generally entertains, not hostility, but the liveliest feelings of gratitude, toward those who have been instrumental in the restoration of reason.

moral effect of a residence in a well regulated asylum for
a time, at the onset of insanity, has an immense effect on
the mind of a patient, and may prevent consequences
that might prove most disastrous were he to be at home
and exposed to the many causes of excitement from
which he is sheltered in an asylum. We must also
decide what form of insanity the patient is laboring
under, and in a medico-legal case must give our diagnosis
as to the insanity of the patient in relation to his civil
capacity and responsibility for criminal action, and also
as to feigned and concealed insanity. In the latter class
of cases—medico-legal cases—it is of the utmost importance for every physician to understand that a man is not
irresponsible for crimes which he commits, from the fact
that some of his ancestors have been insane. The question to be determined here is, whether the hereditary
taint, by being transmitted to the individual in question,
has influenced or determined his volitions, impulses,
or acts. If, on the one hand, he has been noticed for
displaying such peculiarities as usually proceed from
hereditary taint, and if the crime was apparently unaccompanied by any adequate incentive, doubts of his legal
guilt are then to be carefully considered. On the other
hand, if the criminal act appears to have been rationally
performed, and with some adequate and usual incentive,
and if the individual has previously been free from
mental infirmities or peculiarities that might be attributed to hereditary transmission, then we cannot justly
advance insanity as a plea for defense from the consequences of crime. Mental unsoundness, if unconnected
with the testamentary disposition, should not destroy
testamentary capacity. If the will is not affected by, or
is not the product of insane delusion; if the testator has
not ignored the claims of near relationship or of natural
affection; and if his mental faculties are so far normal
that he understands the nature of the act and the consequences arising from it; and if he has a clear idea as to
the amount of property he is disposing of; and if in

making the will he has not manifested any insane suspicion or aversion, the will should be regarded as valid.

The diagnosis of insanity is at times very easily made.
Thus, if we find our patient, from having been previously
moral, affectionate and industrious, has become immoral
and dissolute, exhibits alienation of affections and neg-
lects his business, all without adequate cause, it is of
course easy to determine his insanity, although of course
changes may take place in the character of individuals
without any suspicion of insanity being excited. A
great many cases, however, are on the border line which
separates sanity from insanity, and it often requires the
nicest discrimination to determine whether such a patient
shall be placed under treatment or not.

It now remains to consider the diagnosis of the different forms of insanity which we meet with. In mania the
physiognomy is generally distinctive. The countenance
is furrowed, the eye wild and vacant, and there is generally a peculiar want of agreement in the expression of
the features. The hair often becomes harsh and brist-
ling, and the ears may become shrivelled. The actions,
demeanor, and dress of an insane patient are generally
indicative of mental peculiarities, and oftentimes the
latter may be indicative of the nature of the patient's
delusions, or, if not, it may display marked eccentricity.

In *acute mania*, it is generally easy to discern in the
countenance the presence of some strong emotional characteristics, such as pride, hatred, or anger. It has been
remarked that insanity anticipates the effects of years
and prematurely imprints upon the countenance the
facial lines characteristic of habitual emotions, while in
lunatics of advanced age, these are observable in a greater
degree, and are more deeply marked than they ever are
in sane persons. In this form of insanity — acute mania
—the bowels are generally constipated, the urine is loaded
with phosphates, and the patient suffers from protracted
loss of sleep, which is diagnostic of acute mania, and
which is a symptom that cannot be feigned by an

imposter. Patients of this class pass several days without sleep, and sometimes weeks with but a few hours of sleep in the course of the whole time. Hullucinations of sight and hearing are far more frequent in this than in any other form of insanity. There may be also rapidly changing delusions, and there is generally an intense muscular restlessness, which manifests itself either in destructive impulses, or in continual motion, which rapidly induces dangerous exhaustion, if not properly treated.

In *melancholia*, the most noticeable symptons will be despondency, fear and despair, and the expression of the mental states are depicted in an unnatural degree of intensity upon the countenance of the patient. The patient generally wishes to be alone, is gloomy and depressed, has delusions of fear and persecution, imagines he has committed unpardonable sins, and in the acute cases of melancholia no more pitiable spectacle can be imagined, and the expression of terrible apprehension and fear which occupies the countenance is not easily forgotten. The skin is generally dry, harsh, and muddy, and the bowels constipated. It is such cases as these which have to be carefully watched lest they give way to the suicidal tendencies which are generally present.

In *dementia*, the lines of expression are more or less obliterated, and the vacant, meaningless expression and smile or laugh are indicative of this form of insanity. When the mind is tested, the power of memory, attention, and comparison will be found to be partially or entirely wanting. It is only in primary dementia that the practitioner will find difficulty in reaching a decision, and sometimes these cases are very difficult to determine. In such cases one of the most valuable symptoms is loss of memory. The patient may, in his conduct and conversation, exhibit no marked peculiarities, but when the powers of his mind are tested as to the recollection of past events, or even as to the conversation of a few minutes previous, it will be found that he has entirely forgotten these things. This form of insanity is generally unaccompanied by hullucinations or delusions, and is nearly always due to some exciting cause, such as injuries to the head, attacks of apoplexy, or strong emotional disturbances. There is another variety of dementia which is secondary to acute attacks of insanity, and which differs somewhat from primary dementia. In this form of dementia we meet with the remains of the delusions of acute mania, and we also find an exaggerated state of emotional feeling which remains after the storms of acute mania have blown over, and the functions of the mind are beginning to suffer decay. The diagnosis of *general paralysis* is very easy after we have become acquainted with the disease. In the early stage the most marked symptom is a thickness of articulation, particularly noticeable when the words articulated by the patient are composed of several consonants, when these will be shuffled over in a very characteristic manner. The lips of the patient while he is speaking will be seen to tremble, and likewise the tongue, if it is protruded from the mouth. The gait of these patients is very characteristic and peculiar. They shuffle along in a manner that denotes at once the want of co-ordination in the muscles of the limbs. Later in this form of insanity the power over the sphincters is lost, the patient has to be cared for like an infant, and becomes a great trouble to his attendants.

There is another class of patients whose only manifestation of insanity consists in an abnormal condition of the *moral power*, and who exhibit no obvious intellectual aberration or impairment. The symptoms of the mental disease in these cases are limited to the exhibition of morbid impulses, which the intellect seems powerless to control. These cases of *moral insanity* are sometimes difficult to distinguish, and the laity generally attribute such manifestations to total depravity. In such cases, we must compare the patient with himself when in a state of health and not with any imaginary

standard of sanity or insanity. We should bear in mind
in this class of cases the excellent definition of Dr.
Combe, who says: "It is the prolonged departure,
without any adequate external cause, from the state of
feeling and modes of thinking usual to the individual
when in health, that is the true feature of disorder of
the mind."

We have thus far considered the diagnosis of insanity
only in its relation to the existence of the disease. Let
us finally look at the *diagnosis of recovery*, which
oftentimes becomes a very delicate and difficult task for
the examiner.* We are to determine whether the
patient has recovered so far as to leave no trace of insane
ideas and delusions. We must compare the man with
his former self in a measure, and see if his natural
tastes, affections, impulses, and mental powers have
been restored. Of course we must make an allowance
for a certain amount of weakness in his intellectual
functions, just as we expect to find a man weak bodily
after an attack of typhoid fever or other severe disease.
We must determine whether the man's intellectual
faculties, his memory, reason, and judgment are in a
state to enable him to take his place and position in
active life. We must observe also whether his conduct
is reasonable and quiet. In homicidal or suicidal cases
we must assure ourselves of the disappearance of the
propensity. There are many patients who, although not
recovered, are in such possession of their intellectual
faculties as to become very impatient of restraint and

---

* Dr. Ray says a beginner in this department of our art hails every
improvement as the commencement of convalescence, and is apt to
regard the appearance of a few healthy traits as the unquestionable
presage of recovery. It is not until a later period that he becomes
acquainted with that peculiar oscillation which marks the movements
of mental disease and fully comprehends the fact that serious disorder
may exist in connection with many sound, healthy manifestations of
character. A person may be unequivocally insane, retaining some
flagrant delusion, and yet be calm and apparently rational. With this
exception his views are correct and clear.

confinement, and no amount of reasoning can make them appreciate the necessity for further detention in an asylum. A marked case of this character was formerly under my care, and illustrated forcibly this class of patients, who, if exposed to the excitement of society before a thorough cure has been effected, would almost inevitably have a relapse. This patient would argue for an hour at a time very sensibly and forcibly upon the injustice and oppression of keeping him longer as a patient, and would challenge any proof of his insanity, and probably nine out of ten physicians not acquainted with him would have said that the man was sane. He would converse rationally upon all subjects until the subject of religion was introduced, when he would immediately reveal gross delusions, and would maintain with the utmost sincerity that he could perform miracles, and that he was frequently the subject of them. This shows the importance of examining a patient upon all conceivable topics before pronouncing him cured. These are the cases that generally make their friends and relatives, and particularly strangers, feel that they are unjustly detained, and are the ones who, if they obtain their release in any way, publish their wrongs, and create in this way ill-founded prejudices against institutions for the care of the insane. Generally speaking, if a person who has been insane expresses himself as having been unjustly treated and detained, and denies the fact of his insanity, we may be pretty sure that he has not fully recovered, as persons who are really convalescent are generally fully convinced that they have been insane, and are generally very grateful for the care and attention that have been bestowed on them, and express themselves so. Such patients are nearly always willing to be guided by their physician's opinion as to the proper time for their discharge, and do not, as a rule, exhibit that intense restlessness and desire to return home which is so apt to characterize doubtful recoveries. The first symptoms of recovery are the return of natural tastes, inclinations, and affections in the patient. Drs. Bucknill and Tuke, in speaking of symptoms of recovery, lay down the following excellent rules as evidences of restoration of the mind :

1. A natural and healthy state of the emotions.

2. The absence of insane ideas or delusions.

3. The possession of sufficient power of attention, memory, and judgment to enable the individual to take his part as a free member of society.

4. Tranquil and reasonable conduct; and say regarding them : "When these four symptoms of recovery co-exist there can be no doubt that recovery has taken place." *

PROGNOSIS.

The chances for cure are much greater in recent than in chronic cases. When treatment is delayed the patient's chances diminish greatly, and when treatment is delayed for twelve months, not more than twelve per cent. generally recover their mental health. On the other hand, statistics show that, when the disease is treated promptly, about fifty per cent. may be cured. The results of treatment in cases of insanity resulting from sexual vice are very unsatisfactory, the disease tending toward dementia rapidly. Doubtless a certain per centage of cases relapse, and there is a greater tendency of hereditary insanity to relapse than in any other forms, hereditary predisposition being very unfavorable to permanent recovery, although you may get good re-

---

* It is not safe to discharge a patient while he continues to believe in the reality of any single notion or occurrence that was entirely the offspring of fancy, because such a belief indicates morbid action, which, however circumscribed at present, is ever liable to spread, and induce farther mental disorder. Indeed, the evil is seldom so limited as it seems to a casual observer. A very marked remission, amounting, perhaps, to a complete disappearance of every trait of disease, occurs within the first month of an attack, and is often followed by a renewal of the disease. *This is the result in by far the greatest number of cases.*

sults and cures at first. The influence of epilepsy is very unfavorable, and of course idiocy and imbecility present an unfavorable prognosis. General paralysis is, prehaps, the most unfavorable form of insanity, and is very fatal, cases generally dying in about three years, although death may occur at a much earlier and also at a much later period of the disease. Dementia, with the exception of primary dementia, is also regarded as incurable. Delusional insanity and halucinations and illusions of the senses are rather unfavorable than otherwise, as regards prognosis. Acute mania is a favorable form of insanity and is recovered from, and also acute melancholia. Climacteric insanity presents usually an unfavorable prognosis. Hysterical insanity is very curable. Puerperal insanity is also very curable if seen at once, and a full mercurial cathartic given to commence the treatment, as the cause of the disease is septicæmia, caused generally by absorption of retained products of the placenta. Post-febrile insanity is not very favorable as regards prognosis. Successive attacks diminish chances for ultimate recovery, although you may have repeated relapses and recoveries ensuing. The prognostic value of difference of the pupils in insanity is not great, according to most authorities, and does not seem to justify an unfavorable prognosis; paralytic cases are excluded in these remarks. Cleanliness, restored affections, return to ordinary tastes or habits, are very favorable symptoms, and also the return of suspended secretions. Prolonged insomnia is an unfavorable symptom. Insanity occurring in the young some time before puberty, I believe to be very unfavorable and to tend to imbecility. Respecting menstruation in women, the function may be restored without any corresponding improvement in mind, or the mind may be restored and the menses remain suppressed. Dr. Ray says, of the return of the menses that "we may certainly regard it as a ground of hope in reserve."

Dr. Blandford, of England, in writing on the prognosis of insanity, says:

1. " The general prognosis of insanity will depend on the duration of the existing disorder. Perhaps the best established fact of all is, that the chances of recovery diminish in direct proportion to the duration of the malady, and that it is, consequently, of the utmost importance to place a patient early under adequate and appropriate treatment. If a twelvemonth elapses without appreciable improvement, the chances are decidedly unfavorable. If delusions or hallucinations remain fixed and unchanged at the end of a year, especially if there be hallucinations of hearing, the prognosis is bad. The chief exception is where there is marked melancholia. Patients will recover from this after long periods; whereas such recoveries are seldom found in insanity when depression is absent.

2. "When the cause of the insanity has been of long duration, the prognosis is less favorable than when it is a passing or accidental form.

3. "Is the prognosis unfavorable in hereditary insanity? So much of the so-called simple insanity is hereditary, that we must admit that recoveries from it are not infrequent, for it is from this simple insanity that recoveries chiefly take place. Hereditary insanity is brought about by very slight causes, and thus the prognosis is often favorable, and recovery takes place; but relapse is to be feared, and the prognosis in the second or third attack is not nearly so good. In this hereditary insanity, too, we frequently meet with the cases of recurring and 'circular' insanity, the progress of which is most unfavorable. Both Ray and Greisinger have remarked that the prognosis in hereditary insanity is favorable only where the individual has previously been of normal mind. When he has always been eccentric or semi-insane and undoubted insanity at last manifests itself, the prognosis is very bad.

4. "The more acute the symptoms, the greater the

cerebral disturbance and insomnia, the more favorable the prognosis, if the case is recent. Conversely, the prognosis is bad when there is little bodily disturbance, where sleep is present, the appetite normal, and the secretions unaffected, especially if persistent delusions or an entire moral change are found.

5. "As all deviation from the ordinary mental state and disposition is indicative of insanity, so any return to it is a favorable sign, however trifling the circumstance may be.

6. "Improvement, however slow, is a good sign if it be progressive. So long as this goes on, recovery may take place; but many patients improve up to a certain point and then go no farther.

7. "The age of the patient must be considered. Young people recover in greater numbers than those advanced in life. The latter recover if their insanity be melancholia; but if it be mania with hallucinations and delusions, and obscene conduct and ideas, recovery is rare, especially if the memory is impaired and signs of approaching dementia are present.

8. "All periodicity in the disease, such as exacerbation and remissions on alternate days, is unfavorable."

Drs. Bucknill and Tuke in their manual of "Psychological Medicine" say, respecting the diagnosis of insanity, that no disease is so varied in its manifestations as insanity. That in no other disease do we meet with such an infinite variety of light and shade belonging to their own nature, or to their intermixture with other maladies, or to the influence of temperment, of individual peculiarities of habit, or of social position, and that, therefore, the diagnosis of no other class of diseases taxes nearly so much the ingenuity and patience of the physician. The physician is compelled to bring to the investigation of mental disorders, a clear analytical conception of those functions which collectively constitute mind.

The diagnosis of insanity presents itself either in a strictly medical or in a medico-legal point of view. If the question is of the former character, not alone the kind of medical treatment, but also, the question as to whether the patient has to be deprived of his liberty comes up. If the question is medico-legal, we may have to appear either in civil suits and proceedings or in criminal trials. In civil suits the distribution of property to a vast amount, the validity of wills, contracts and other social and commercial acts, often depend upon the decision of the physician; and in criminal trials the frequent issue of the question is the awful one, whether a human life shall be sacrificed with violence and ignominy, or spared by establishing the plea of not guilty, on the ground of insanity. Whether the question be purely medical or medico-legal in its bearings and apparent consequences, the grounds of the diagnosis must be the same, for, although in criminal trials the nature of the crime itself, and the manner in which it has been effected, must often be allowed to have no inconsiderable weight in the formation of the judgment, yet, these circumstances are essentially no other than a part of the conduct of the patient; and the conduct must be carefully estimated, even when the question is most purely medical. The physician is called to see a patient whose symptoms have caused alarm and anxiety to his friends. They wish to insure both his safety and their own, and to provide immediately the treatment which affords the best promise of recovery, and above all, to have the momentous question decided for them of confinement in an asylum or of treatment at home.

The diagnostic value of hereditary tendency is great. The insanity of one parent indicates a less degree of predisposition than that of a parent and an uncle, and still less than that of a parent and a grandparent, or of two parents. The insanity of a parent and a grandparent with an uncle or aunt in the same line, may be held to indicate even stronger predisposition than the insanity of both parents. The influence of the insanity of parents in creating a predisposition will depend, to a great extent, upon whether it has taken place before or after the state of parentage commenced. The insanity of a parent occurring after the birth of a child, if it arose from a cause adequate to excite it without previous predisposition, would, of course, be held as of no value in the formation of hereditary tendency. The insanity of brothers and sisters may be of much or of little value as evidence of predisposition, according to the circumstances under which it has shown itself. If several of them, both older and younger than the patient, have become insane, the fact tells strongly in favor of predisposition, although neither parent nor grandparent may have been insane; since it is well known that other conditions in the parent, besides that of actual insanity, may create this predisposition ; for instance, violent and habitual passion, the debility of old age, and most of all, habits of intemperance at the time of procreation.

The diagnostic value of previous attacks is considerable, as few diseases more frequently recur than those which affect the mental functions of the brain. A slight and transient attack, however, respecting the real nature of which there may have been some difference of opinion, will be of very different import from a prolonged attack of decided character. The greater the length of time which has elapsed since any previous attack has been recovered from, the less will be the value of it as an indication of the nature of the existing disease.

The diagnostic value of change of habit and disposition is very decided. A comparison of the present behavior and habits of his patient with those which existed in a state of health often will afford the physician a most satisfactory evidence of morbid change in the brain. The natural character of a man who is insane is either changed or exaggerated. The vagaries of hysteria in a woman must not be mistaken for actual insanity. The physician may see in his patient one of four things : first, a vacant and meaningless expression, and a childish absurdity of action, the signs of dementia, of imbecility, or those of general paralysis ; secondly, a facial expression of deep and concentrated sorrow ; or thirdly, indications in physiognomy, or demeanor, of strangeness and irregularity ; or fourthly, no outward indication of mental disease.

In melancholia the patient will readily converse on his mental symptoms. In imbecility and early dementia his apprehension is not sufficiently alert to place him on his guard ; and in mania, he either suffers from head symptoms, respecting which he will readily talk with the physician, or his mind is actively engaged on some project or object, which will afford the physician appropriate topics for conversation. The most difficult cases are those in which differences of opinion and of interest exist among the members of the patient's family, and the patient has been quietly told that it is wished to prove him insane and to place him under confinement, and that the doctor is coming to examine him for that purpose.

There is often a diagnostic value in peculiarites of residence and dress. The author had a case of general paralysis brought to him for diagnosis. The patient, a man of wealth, had *three* handsome neck-scarfs on and *several* valuable scarf-pins, and informed us of his desire to send us *one thousand boxes of cigars* as a present. Said he felt *magnificently*, that there were few stronger men in New York than himself. At that time he was advanced some thirteen months in the course of this intractable malady, and had the shuffling gait and the diagnostic hesitating stammering speech of a general paralytic. He was full of delusions of wealth and grandeur.

The diagnostic value of peculiarites of bodily condition may be important. There may be emaciation from loss of rest, derangement of the alimentary processes, a quicker pulse than normal, and a tongue coated in the centre. The skin as a rule is harsh and dry and the complexion muddy. We frequently find disordered states of the

9

abdominal viscera in insanity, and we may not unlikely discover gastric or hepatic disorder. Uterine disease is very frequently present. The outward expression in the patient's features and gestures of his inward psychical state of sadness, melancholy, despondency of despair, may be very striking at times. Intensified expressions are seen in insanity of the various emotions, such as pride, anger, fear, jealousy, and the patient with partial insanity may exhibit an unvarying and intense expression of any one particular emotion. In mania the attitude is restless, the motions quick and expressive of various and changeful emotions, while in melancholia the attitute is apt to be fixed and the gestures slow. In imbecility and dementia we see slovenly postures and undecided and aimless movements.

Respecting the *physiognomy of insanity*, the extreme distortion of the features produced by acute mania or acute melancholia, is unmistakable. There is a much greater expression of intense pain in cerebral inflammation, attended by maniacal symptoms, and a more prominent bloodshot eye, than in mania alone. In the delirium of fever the countenance indicates low emotional force, while in the delirium of mania the facial expression of emotional force is highly exaggerated. If there is mobility of the facial muscles in the delirium of fever it is tremulous and feeble, indicating want of power, while in mania the play of these muscles is full of expression and power. It is vigorous and tense, indicating a concentration of nervous force. The wrinkles in the delirium of fever are the result of emaciation, while in the face of the insane man they are caused by the tense contraction of the muscles of expression. There is an apparently causeless and motionless play of features often seen in the insane.

In melancholia the facial expression is emotional. In mania it is emotional and intellectual, and marked by the above characteristics of changeableness and inconsistency. In dementia all expression has disappeared, and the physiognomy is vacant and meaningless, showing an absence of thought and desire. General paralytics exhibit trembling lips, drooping brows, and features expressive of a mingled state of imbecility and excitement, eyes with pupils of unequal size, all of which constitute a *tout ensemble* perfectly diagnostic to the experienced alienist. In primary dementia it is sometimes difficult to make a decided diagnosis. The demeanor and conduct are very slightly changed, there is nothing strange in the appearance, but a great diagnostic sign is loss of memory for very recent events. In conversing with the patient he may not be able to remember what he has been talking about a few minutes previously. Injuries to the head and apoplexy most often cause it, and fever and emotional disturbances, especially grief, will also cause it. There is absence of delusion or hallucination. The physiognomy may be silly and meaningless, and the eyes may have a meaningless look, and there may be a vacant smile on the lips.

The patient may also lose not merely the power of understanding anything like an intricate account, but the value of very simple numbers. These facts account for the reckless expenditures of patients with recent insanity.

Acute and chronic mania, and also *incomplete mania*, may be easily recognized, or in the latter case the diagnosis may be extremely difficult, and we may have absolutely nothing but uncontrolled propensities and extraordinary conduct to guide us.

In *chronic mania*, especially with lucid intervals, we may find a remarkable strength of all intellectual functions, in so far as they are not affected by delusions. The perceptive faculties are retained in all their activity, and the memory is very good, and even the judgment, on matters unconnected with the delusive opinions and perverted emotions peculiar to the case, may not be greatly affected. The delusions may be numerous or they may be few. There is grave emotional perversion.

In *incomplete primary mania* there may be a decidedly abnormal state of the emotions and sentiments without marked intellectual lesion. This symptom is constant. Friends and relatives are detested and abused, and the objects of natural affection overwhelmed with invective, and, perhaps, sacred things made the subject of blasphemy. This moral perversion clearly indicates insanity, but there are slighter shades of perverted emotion which require all the adroitness of the experienced alienist to discover. Absurd opinions are generally allied to perverted emotions. Exaggerated hysteria may confuse the diagnosis, and it may be mistaken for incomplete primary mania, but the age, sex, constitution and character of the patient will generally reveal the nature of hysterical attacks when they occur. I had an hysterical patient who feigned that she had the delusion that there was an animal in her abdominal cavity, and this was in strict keeping with the tenor of her life, for she feigned everything; she, however, made a beautiful recovery, by the use of the wire brush electrode, with the strongest induced current. In making the mental examination we test the fundamental faculties, the attention, the memory, and the judgment, and lead the patient to give an account of his own powers of body and mind with reference to health, to exercise, diet and study. Thousands of delusions are entertained by insane people upon these subjects. A conversation respecting the patient's possessions, his means of livelihood, and his hopes of advancement, will lead up to delusions of pride, ambition, and acquisitiveness, if such exist; carrying the conversation on to his near relatives, and friends, birth and parentage, and the patient's belief whether his parents were his actual and real parents, will lead up to delusions respecting imaginary greatness, and any perverted emotions towards those who ought to be dear to him. His religious observances may be inquired into, with the expectation of finding insane delusions on this subject. Politics and science may be made the topic of conversation with an educated man, and, if insane, he will hardly stand the test of discriminating inquiry on these and similar subjects. Indecorous conduct towards the opposite sex, perverted appetite and unnatural habits, we must learn of from those who have opportunities to discern them.

The diagnosis of eccentricity is only likely to be brought up in cases of disputed wills or in criminal cases where eccentric conduct is utilized to support the plea of insanity. There are two forms of eccentricity. The one arising from an excess of individuality, where the individual is often endowed with more than an average portion of good sense and of moral courage, although his sense is founded upon reasons marked out by his own mind, upon propositions laid down by himself, and adverse to the common sense of those among whom his lot is cast, and his moral courage is displayed by adhesion to his own opinions, and by setting at naught the ill-founded ridicule of the world. An eccentric man of this type is further removed from the chances of insanity than most of the sane people upon whose prejudices and fancies he sets his heel. His intelligence is not made the sport of his passions, his emotions are under control ; in short, he has superior intelligence.

In the second form of eccentricity the man deviates from the ordinary observances of society, from weakness of judgment, from love of applause, and the desire of drawing upon himself the attention of others. His conduct is ill-regulated and influenced only by vacillating emotions, strong or weak, according to the caprice of the hour. He has intellectual powers of low order, great desire of approbation, and little individuality. This form of eccentricity is often nearly allied to insanity and is often premonitory to it. Its subjects are to be found in families tainted with hereditary predisposition to mental disease, and it merges so gradually and insensibly into mental disease that the lines of demarcation are traceable only with the greatest difficulty, and, indeed, often are not to be traced at all. In many cases, however, the transition is marked by perversion of the emotions, by unfounded suspicions, anxieties and antipathies, and also by signs of physical disturbance, by sleeplessness and general feverishness.

The diagnostic symptoms of *melancholia* are despondency, fear, and despair, existing in a degree far beyond the intensity in which these emotions usually affect the sane mind, even under circumstances most capable of producing them, and in numerous instances existing without any commensurate moral cause and often without any moral cause whatever. The sad and anxious eye, the drooping brow, the painful mouth, the attenuated and careworn features, the muddy complexion and harsh skin, the inertia of body, the stooping and crouching postures, the slow and heavy movements, speak of distressing oppression of the faculties and intense wretchedness. In other cases fearful anxiety is observed, and the eye becomes bright, the nostrils dilated, the movements quick, irritable, and often impassioned under the influence of some vague terror. If the physician can note the above symptoms and can trace them to a cause productive of insanity, he will have little difficulty in pronouncing his patient insane, although he can discover no trace of delusion. In many cases the patient is painfully aware of the nature of his malady, and seldom attempts to conceal his consciousness of it from any considerate and sympathizing inquirer. Generally, in melancholia, there are intellectual errors displaying themselves by false sensation, perception, or conception; in illusion, hallucination, or delusion proper. There is first emotional and secondly intellectual disturbance in melancholia. Respecting the differential diagnosis between hypochondriasis and melancholia, Prichard said "that a hypochondriac is in full possession of his reason, though his sufferings are not so dangerous or so severe as he supposes; but if he declares that his head or his nose has become too large to pass through a doorway, or displays any other hallucination, he has become a lunatic; his disorder has changed its nature, and this conversion takes place occasionally, though by no means so frequently as is supposed." The apprehensions of the hypochondriac are confined chiefly to his own feelings and bodily health. On other subjects they converse cheerfully, rationally, and justly, while melancholiacs view all things through a gloomy medium. The cause of hypochondriasis is generally dyspepsia or some morbid state of the digestive organs. The love of life and fear of death characterize hypochondriasis, while a frequent symptom of melancholia is disgust of life, attended with desire to commit suicide, which, when motiveless, is one of the surest marks of insanity.

The diagnosis of *monomania* is easy, from the prominence of the single intellectual error. The great majority of cases are sequences of or transformations from melancholia. The emotional disturbance comes first; the intellectual afterwards.

The diagnosis of *moral or emotional insanity proper* is sometimes very difficult. This is, according to Blandford, a disorder of mind shown by an entire change of character and habits, by extraordinary conduct and acts, extravagance or parsimony, false assertion and false views respecting those nearest and dearest, but without absolute delusion. It may follow epileptic or apoplectic seizures, or may be seen after a period of drinking. Its approach is gradual, as a rule, rather than sudden, and the extraordinary character of the acts may not at first be so marked as subsequently. Friends wonder that a man should say this or that, or should do things so foreign to his nature and habits, but some time may elapse before they can convince themselves that such conduct is the result of disease; and the acts may be such that many will look upon them, even to the last, as signs merely of depravity. Such insanity, of course, varies in degree. When it is well-marked and the conduct is outrageous, there will be no difficulty in the diagnosis. But it may be less marked. It may consist of false and malevolent assertions concerning people, even the nearest; of little plots and traps to annoy others, in which great ingenuity and cunning may be displayed ; and there will be the greatest plausibility in the story by which all such acts and all other acts will be explained away and excused. It would seem sometimes as if a universal badness had taken possession of the individual, yet a badness so inexplicable that it can only be looked upon as madness. Much examination and opportunity for examination may be needful before we can sign a certificate, for such people are often very acute and quite on the alert. They have no scruples about falsehood, and will deny or justify everything with which they are taxed. And where the insanity is manifested in conduct, the medical man may never be a witness of it, and is obliged to receive on hearsay that which the patient strenuously denies. Careful inquiry, however, will probably reveal the origin and cause of the change ; there may have been a period, though short, of acute insanity,—as acute mania or melancholia,—which passed away and left this as a permanent condition ; or it may be the precursor of a more advanced stage of insanity marked by the ordinary symptoms of delusion and hallucination. If the change has been rapid and progressive, and more and more outrageous and eccentric, it is likely that in a short time unmistakable insanity will be displayed. The one constant and marked feature of this insanity is the absence of delusion, but we are not, on this account, to argue that the intellect is sound.

The hardest form of moral insanity to estimate and diagnose is the *congenital moral defect*—the *moral imbecility* occasionally met with in cases of this reasoning mania. These patients are utterly incapable of telling the truth or of understanding why they should do so. These are the cases that commit crimes and are very dangerous to the community in which they live. They may have considerable intellectual ability.

The diagnosis of *general paralysis* is easy to one familiar with the disease. The best symptom for early diagnosis is the modification of the articulation. It resembles the thickness of speech in a drunken man, and depends upon loss of power over the co-ordinate action of the muscles of vocal articulation. Words composed of numerous consonants, with few vocalic sounds, are articulated in a shuffled manner that is perfectly characteristic. In speaking, the lips are tremulous, as if the patient were about to burst into tears. Protrusion of the tongue is difficult, and it cannot be long protruded, and while protruded it quivers. The brows droop, and the contraction of the iris, under the stimulus of light, is often different in the two eyes. The voice has a peculiar tremor, and the gait is stumbling and shuffling. Later on the power over the sphincters is lost, and finally the patient may choke to death by the stoppage of food in the pharynx. The psychical symptoms are generally delusions of wealth and grandeur. There is a universal extravagance of ideas. There is loss of excito-motor sensibility.

The detection of *feigned insanity* is very important. Those who feign insanity generally overact their part. The long-continued sleeplessness of mania cannot be feigned. Neither can the restless, continued agitation, the rapid pulse, the foul tongue, the dry, harsh, inelastic skin. If the skin feels healthy and sweaty from the exertion of the pretender, and if he sleeps soundly and composedly, we may be pretty sure he is feigning. Chronic mania is more easily simulated and more difficult of detection. Respecting these cases, that before deciding upon the reality of any doubtful case of insanity, all the physical conditions of the individual, such as the amount of sleep, the state of the pulse, skin, tongue, and digestive system generally, the conduct and the state of health immediately preceding the signs of insanity, should be ascertained.* The effect of remarks made within hearing of the suspected person should be observed. One who proclaims his own insanity should be distrusted.†

The diagnosis of *concealed insanity* may, at times, be made by inducing a patient to write to some friend, when things that he would not speak of he may write of at some length, and his delusion be made very apparent. A patient's conduct should be watched by night as well as by day to discover concealed insanity.

---

* True insanity is generally hereditary and preceded by well-marked premonitory symptoms, or, if sudden, has an adequate or sufficient cause, such as accident, disease, loss of friends or fortune, etc. The criminal lunatic rarely seeks to escape ; the imposter generally seeks to escape.

† If the simulator refuses to answer all questions, refuses food, has a stupid expression of face, and remains obstinately silent, it may be at times difficult to detect the simulation. Bucknill and Tuke say that the most important diagnostic mark of feigned insanity is a want of coherence in its manifestations—their unconformity, not only with mental disease in general, but with the form or variety of insanity which is feigned in particular. The simulator mixes the forms of insanity together.

CHAPTER VII.

CIVIL INCAPACITY; TESTAMENTARY CAPACITY; LEGAL
TESTS OF RESPONSIBILITY; HINTS FOR GIVING TES-
TIMONY; EXPERT TESTIMONY, AND THE FUNCTIONS
OF EXPERTS IN INSANITY.

*Civil Incapacity; Legal Tests of Responsibility;
Hints in Giving Evidence.* Respecting the civil incapacity of an alleged insane man, *the acts of any person,
either in or out of an asylum, may be declared invalid,
if it can be shown that, at the time they were performed,
the person labored under such an insanity as rendered
him incapable of performing them rationally and
without injurious consequences.* On this principle any
person may be found to have been incapable of contracting marriage, of executing a deed, contracting a debt,
making a will, or giving credible evidence. The principle, it must be carefully noted, is not that the mere
existence of insanity in the person performing them
invalidates such actions, but that if the insanity has
materially affected the character and quality of the
actions, they may be thereby invalidated. This is one
of the most important principles that a medical jurist
has to keep in mind, as it is not an unfrequent mistake
to suppose that a person is necessarily incapacitated for
the performance of every civil act the moment he can be
proved to labor under any condition to which the term
insanity may be applied. Perhaps the case in which the
validity of a civil act is most easily endangered by the
existence of any form of insanity is the contract of

marriage. This proceeding is supposed to so affect all the relations of life that almost any form of unsoundness of mind may be sufficient to interfere with that intelligent and deliberate consideration which is essential to the giving of rational consent. In these cases—medico-legal cases—it is chiefly important that the practitioner should distinguish : 1st. Diseased perversion of the mental faculties. This includes all kinds of insanity which are the result of active disease, such as the simple form of delirium, mania, melancholia, and monomania. 2nd. Weakness or enfeeblement of the mental faculties, resulting either from defective development, disease or decay. This includes congenital imbecility, and all the forms of what is called chronic dementia, all those enfeeblements of mind which are sometimes the remaining effects of acute disease, sometimes the concomitants of chronic disease, and sometimes only the mental phase of senile decay. In order to establish the incapacity of a person said to labor under any of these forms of disease, it must be necessary that an experienced physician should not only be able to detect their characteristic symptoms, but also to show that the performance of the duties or the exercise of the rights under consideration would be modified or obstructed by the existence of such disease.

Marriage. As has been already stated, the mere existence of any form of insanity in one of the parties may render a marriage contract void.

Respecting insanity as a plea for divorce or nullity, we think that divorce should be permissible : 1. From a woman who, subject to recurring attacks of insanity, marries during one of the periods of health or lucid intervals. 2. When it can be shown that one of the parties has concealed the fact of previous attacks of insanity having occurred. 3. When either of the contracting parties, at the time of marraige, was incapable of understanding the nature of the contract.

We cannot overestimate the importance of the nature of the earliest symptoms of the insanity, and the precise period when they appeared, and when they so far affected the mind as to render the individual irresponsible.

It is not rare to see cases of women who at certain times develop erotic feeling which may be followed by other signs of mental disorder, and during this period of disturbance a woman may disregard all her social and other ties and enter into a foolish marriage of which she bitterly repents as soon as her mental tone has been recovered. In such cases divorce or nullity might properly be granted.

Women of intellect and beauty who break off engagements on trifling pleas, as a rule have some temporary mental or bodily disease which, when cured or disappeared, leaves them much disappointed and chagrined with their past action. There are a class of unstable, neurotic girls, with little or no sexual desire, whose icy indifference is often wrongly attributed by lovers to coyness and shyness when engaged, and who, when married, will decline to fulfil the functions of a bride and allow her marriage to be consummated. These are cases who easily pass the border land into actual insanity. This was the case in the English case of Hunter v. Hunter or Edney, where the husband petitioned for nullity on the ground that his wife was insane at the time of marriage. The judge decided that the woman was not in a position to enter into the contract in consequence of being of unsound mind, and declared the marriage null.

In the case of Durham v. Durham in England, the marraige was consummated, and soon after symptoms of insanity appeared, and there was dullness before marriage and possibly some delusions, but clear evidence was not put in on this point. On this case, marriage probably precipitated the threatened attack of insanity. Sir James Hannen decided that she was sane when married, and dismissed the petition of Lord Durham for divorce or nullity.

In the case of Cannon v. Cannon or Smalley, in England, no signs of insanity were noticed until the day

before the wedding. She became then dull and unhappy
and was absent-minded at the wedding breakfast. She
refused to allow the marriage to be consummated.
Melancholia appeared, succeeded by a lucid interval,
during which the marriage was consummated, and then
the woman exhibited strongly erotic ideas. A prolonged
state of mania followed.

The decision in this case was against the petitioner for
divorce or nullity, the president holding that she was not
of unsound mind at the time of marriage. Medically,
this woman was unquestionably insane when married.

Civil Contracts may be held binding although made
by lunatics. If the person with whom a contract is made
had no knowledge that the person contracting was insane,
and if no attempt was made to take undue advantage of
him, the contract would be held good.*

Testamentary Capacity. A person is considered to be
of a disposing mind, that is, capable of making a valid
will, if he knows the nature of the act which he is performing and is fully aware of its consequences. It is in
regard to the making of wills that the law has carried
out most thoroughly the principle that the validity of an
act ought to be maintained in cases of insanity, unless at
the time the act was performed the state of mind of the
agent can be shown to render him unfit to perform that
particular act in a rational manner. Persons have made
valid wills while inmates of lunatic asylums. And one
will was held to be good, though the testator had committed suicide within three days after its execution.
The existence of delusion, which has been regarded by
lawyers as of such importance in cases of alleged insanity,
does not invalidate a will; for it has been declared to be

---

* The contracts of the insane are, in many cases, declared to be invalid,
and are set aside on the ground of fraud in accordance with an established principle that the parties to a contract must be capable of giving
their deliberate and rational consent, the power of doing which is
destroyed by mental derangement. (Story's Commentaries on Equity
Jurisprudence, § 227.)

"compatible with the retention of the general powers of the faculties of the mind," and to be "insufficient to overthrow the will, unless it was calculated to influence the testator in making it." [We had, recently, under our professional care, a young lady of wealth, a case of chronic mania with lucid intervals, in whom the natural affections were more than usually lively, who possessed a perfectly clear idea of the amount of property she possessed, and the way in which she proposed disposing of it in the event of her death, and whose will, as dictated by her, was as sane a document as we ever examined. Her testamentary capacity, although an incurable case of insanity, was perfectly good, and her will perfectly valid.]

On the other hand, a will may be invalidated on account of the existence of mental states which would not be regarded as insanity from either a legal or medical point of view. Drowsiness and stupor resulting from erysipelas or fever, extreme weakness from cholera and failure of memory in old age, have all been found sufficient to void a will. If a physician is called on to be a witness to a will, it is his duty to satisfy himself as to the testamentary capacity of the testator. His subsequent evidence in regard to this, will, in case of dispute, be of almost decisive influence if he has taken proper means of forming an opinion. In all cases, therefore, where there may be a possibility of doubt, it is well to require the testator to show that without extraneous aid and without referring to the document itself, he remembers and understands all the provisions of the deed.

Dr. Ray says: "In that form of mental disorder called acute mania, in which the mental movements are continously strange, wild and incoherent — madness without method — the testamentary incompetence of the patient has never been questioned. Indeed, the condition is one in which we should hardly expect any disposition to make a will, and I am not aware that such a one has ever been offered for probate. Lord Coke, you observe, speaks of a description of lunatics who have sometimes understanding and sometimes not. This statement has reference to a phenomenon once supposed to be a very common occurrence — I mean that of lucid intervals, as they are called. So common were they thought to be, that in every case of alleged insanity, the question of a lucid interval was always raised as a matter of course. In fact, the practice has not yet entirely ceased. The idea is, that in most insane people reason returns at intervals, and with it their original competence and responsibility. The law not only supposes the probability of such a period, but the party availing himself of the plea of insanity was obliged to show — even though the burden of proof may not always have been put upon him directly — that there was no lucid interval. Considering how small a foundation this whole doctrine of lucid intervals has, it is difficult to account for its prevalence. Unquestionably, it sometimes happens that an insane person comes to himself, manifesting his natural propriety of conduct and conversation, his memory and perception apparently clear, the cloud returning after a few days as dark and dense as before. Exactly how far the mind in this condition is free from the influence of disease we never can know. Considering, however, the suddenness of the change, the brevity of its duration, and the long continuance of the disease when it occurs in chronic cases, it is not likely that the mind is restored to its normal degree of clearness. It must be borne in mind, also, that, as described by distinguished legal authorities, such intervals are of very rare occurrence. I have not seen more than half a dozen cases in all my experience. *All nervous diseases are subject, more or less, to a certain law of periodicity, by reason of which at intervals their regular course is changed,* and other incidents come and go in a certain order of succession. The change thus produced may sometimes amount to an entire disappearance of the signs of disease. This phenomenon is not unfrequently witnessed in the early stage of acute mania. Within the first month there may occur a rather sudden cessation of the manifestations of disease, in which the patient is calm, quiet, talks and behaves sensibly, though, if closely observed, there will be found some indistinctness of memory and confusion of thought, especially with reference to the circumstances of the attack. This condition has often been confounded with lucid intervals, especially by lawyers, who find it difficult to see a distinction which can be visible only to the long-practiced observer. To the common eye, *any remission* in which the patient is tolerably calm after being violent, and answers a few questions rationally, seems like a lucid interval.

" Admitting, as I do, that a valid will may be made in the lucid interval, it is likely that some shadow of disease may rest upon it, that any testamentary act during that period should be very closely scrutinized. Some qualifications for the act are required not otherwise needed. It should be shown that changes in the circum- stances and conditions of those whom the testator is disposed to benefit (having occurred when he was incapable of understanding and appreciating them) were brought to his knowledge and comprehension, since such things would naturally affect the dispositions of the will; because it must be borne in mind that the past, for months or years, may have been a complete blank, or filled with strange and deceptive images. In short, we may conceive of a case where every testamentary qualification was possessed ; but certainly the fact must be of rare occurrence, and difficult of proof. I know of only one case reported of a will made in what was called improperly a lucid interval, and established, that of Cartwright v. Cartwright, 1 Philimore, 90. The testator was in an asylum, and so severe was the grade of her disease, that she had restraint on her limbs at the moment when she called for pen, ink and paper in order to write her will. This she did at last after writing one on several pieces of paper, which she tore up and threw

10

into the grate, and walking up and down wildly, and muttering to herself. The will was established, the court deciding that it was made during a lucid interval on the strength of the internal evidence, as it made a natural and consistent distribution of her property. This fact the court considered conclusive proof that a lucid interval had taken place. The result may have been right; but it was reached by a sort of logic known as reasoning in a circle. The correctness of the will proves the interval, and the interval being proved, makes the will valid. *Had the court, while squarely admitting the insanity of the testator, declared that the character of the act showed that she was still rational enough to make a valid will, it would have uttered good sense and good science.* Here we see the binding influence of the old law as expounded by Coke. It being proved that the testator was a lunatic, she was necessarily *non compos*, and could become otherwise only by recovery or a lucid interval.

"This influence has not yet entirely lost its force, for I observe that lawyers are not content with proving competence sufficient for the act in question, but also labor hard to prove also perfect soundness of mind. In a case that came under notice two or three years since, an attempt was made to void the contracts of a person recently deceased, on the ground of insanity. It was shown that the transaction — the sale of coal lands — was just and fair in every particular; that he obtained a price pronounced by his neighbors and advisers to be a fair one; and that he sold, then and there, without waiting for a prospective rise in value, simply because he needed the money. That he was a lawyer in full practice, a leader at this bar almost up to the day of his death, and all the while the trusted counsellor of several large corporations — all this needed no proof, for it was seen by everybody. And yet, because of the admitted fact that this man had always been remarkably eccentric in his ways and manners, week after week was spent in endeavoring to show either that he was or was not technically insane. If the administration of the law has for its object the promotion of justice among men, we may venture to say that the means it used in this instance were signally irrelevant.

"Farther examination of Coke's classification of the *non compos mentis*, furnishes fresh proof how little help it gave the medical jurists in settling the questions that came before him.

"'One class,' he says, 'consists of those who, by sickness, grief or other accident, have wholly lost their memory and understanding.' The word *wholly* is probably used inadvertently, because is would refer to persons in the very last stages of dementia, whose acts would scarcely become matter of litigation ; whereas he had in mind, no doubt, a numerous class, who, while moving about among men, and taking some part in the affairs of life, are, nevertheless, laboring under considerable mental infirmity. We have reason to think that this class was meant to embrace the subjects of senile dementia, of the weakness of old age, and of the damage inflicted by paralysis and other cerebral affections. His allusion to grief and accidents implies, probably, a mistaken notion he had conceived respecting the cause of the mental affections. Taken as a whole, this class, unquestionably, has always furnished the courts with a greater amount of litigation than all others put together, in the matter of wills. The more exact and well-defined are our notions of mental capacity, the more foundation they have in close personal observation of this class of persons, the better will be our admiration of justice. Much of the apparent conflict in the adjudication of their cases has arisen, I think, from a faulty appreciation of the mental qualities chiefly concerned ; and therefore I invite your attention for a moment to this point.

"Old age is usually accompanied by a certain enfeeblement of the mental as well as the bodily powers. This condition does not imply unsoundness or incompetence. It merely means a diminished power of endurance; an incapacity for those long-sustained efforts once comparatively easy; more difficulty in grasping obscure and remote relations. The mind has lost none of its characteristic tastes, and none of its fondness for its accustomed pursuits; but it is satisfied with easier tasks and welcomes longer intervals of rest. The only mental faculty obviously involved in this condition is the memory, especially of recent incidents, even while old ones are well retained. The forgetfulness of young people comes from carelessness and a predominant interest in other thoughts. With the aged the new impression, however vivid at first, fades from a lack of power in the brain to retain it. It must be quickly and frequently repeated, before it will endure. So far the change may be considered as the normal result of old age, and destitute of any legal consequences. Occasionally, indeed, in the closing years, the mind displays even more than its wonted vigor and brilliancy. It was these exceptional instances that led the poet Waller to say:

> " ' The soul's dark cottage, battered and decayed,
> Lets in new light through chinks which time has made.'

"Sometimes the change here described proceeds still farther, and induces a condition that is abnormal, ending in *senile dementia*. The memory becomes less and less tenacious; the perceptions less exact and clear, one person or thing being mistaken for another, and their relations misunderstood. Then the judgment — that is, the power of discerning the relations of cause and effects, of distinguishing between the specious and true, of taking in the remoter considerations germane to the case in hand, of weighing and rightly appreciating conflicting claims, loses its vigor, and is easily led astray by false lights. And so the process of decay goes on, until it reaches its utmost limit in virtual extinction of the mental powers. Now what we would like to know

is, the precise point at which testamentary capacity
ends ; and this, of course, is beyond our reach. As to
the effect of this condition, both in its earliest and its
latest stage, there can be no diversity of opinion. It is
during the intermediate stage that it gives rise to
obscurity, doubt, conflicting evidence and abundant
litigation.

"To arrive at a correct decision, we must first understand what are the intellectual powers necessary to
testamentary capacity. In the first place, the memory
must be active enough to bring up to mind all those who
have natural claims on the bounty of the testator ; to
make him aware of the nature of his property, its location, the incumbrances upon it and his debts. If he
makes bequests in certain sums of money, he should
know with some degree of exactness the value of his
property ; and if he has made previous wills, he should
be aware of their contents. Whether he actually had
such a memory will generally be determined, for the
most part, by the circumstances of the transaction. In
most cases evidence is given respecting the general condition of the memory as manifested in the ordinary
discourse, and this, coming, as it does, in a loose,
detached, fragmentary manner, from persons usually
unaccustomed to observe mental phenomena closely,
requires to be carefully and intelligently examined.
The *lapses* of memory exhibited by all old people must
be distinguished from that *utter loss* of memory that no
effort can retrieve, even for a moment. The former is
chiefly in regard to recent things, which are readily
brought back to mind and are retained for awhile. The
latter embraces old as well as recent incidents, impressions customary as well as casual, ideas the most as well
as the least familiar. The old man who is constantly
mislaying his spectacles, forgetting the face of the
person to whom he was introduced the day before, and
marching up the broad aisle of the church with an
umbrella over his head, may be found, when his atten-

tion is specially directed to a subject, to remember its prominent points, understand them well, and govern himself accordingly.

" Let it be observed, in this connection, that many of those old people so forgetful of passing events and so careless of little proprieties, need only to have their attention fixed on the matter in hand to display no lack of memory or understanding. If, on the other hand, a person has utterly forgotten the events of his earlier age ; if he cannot tell his own age, or the year of our national independence, is unable to tell how many six and six make, and has forgotten whether his estate is in lands or houses or stocks, he surely has lost his testamentary capacity.

"The mental infirmity most often the source of testamentary incapacity, is impaired judgment. To make an equitable distribution of his estate among those connected with him by blood or affection — such a one I mean, as he would have made while confessedly sound — implies on the part of the testator a variety of considerations that cannot safely be overlooked. He should be able to appreciate properly the nature of their claims, their present and prospective necessities and the favors they have already received ; and all this, not to mention other considerations, requires a nice discrimination, and the power of looking before and after. If the bequests indicate any deficiency in these respects, it certainly furnishes ground of suspicion.

"There is a large class, you observe, still unprovided for in the schedule of Coke. I refer to that of persons neither idiots, lunatics, nor the victims of sickness, grief or old age — persons having by nature a deficient mental endowment, and embraced under the general appellation of imbeciles. The wills of such persons often come into dispute, and though their disposal is determined by the same principles as those last referred to, yet they give rise to a larger range of speculation and doubt. In the one case, the question on which the result may depend is, How much mental power has been lost? while in the other the question is, How much was ever possessed? The difficulty of the question is increased by the fact that in many, if not most imbeciles, there is much inequality in the strength and development of the several mental faculties. The same person may be shrewd, even sharp, in some transactions, dull and foolish in others; at one moment uttering a pithy remark, at another leaving no doubt of his native simplicity. With shrewdness and folly thus displayed side by side, it is not strange that different observers are sometimes very differently impressed by what they witness. One who has sold him a gun or a fishing-rod, or made for him some little article of use or ornament, and listened to his comments, is ready to pronounce him about as sensible as the generality of men. Another, who has witnessed his conduct in affairs requiring some prudence and judgment, is strongly impressed with the depth and breadth of his simplicity. And this is a specimen of the evidence heard in our courts, when they are called to adjudicate in cases involving the persons or property of simple-minded people. It is also a lamentable fact, that the disposition to form positive conclusions on the strength of a partial, one-sided observation, is about as common among men of some culture as it is among those who are without pretentions to any. Until mental manifestations are better understood, we shall continue to witness these strangely conflicting conclusions in litigation involving the interests of imbecile persons.

"To meet this state of things, the first thought was to fix upon some arbitrary, specific standard of mental power, by which the minds of imbeciles should be measured. I know of only one that has survived the test of experience, and even that is seldom offered now. It has been said, with some show of authority, that to make a valid will one must have capacity sufficient to make a contract. Had these conditions been reversed, and the proposition been, that as much mind was required to

make a contract as to make a will, it would have had as
much support in the nature of things as the other, and
that is none at all.   Until we are satisfied as to the exact
amount of mind necessary to make a contract, this
measure will scarcely help us, even with the estimate I
once heard given from the bench, that it is as much as is
required in the ordinary business of life ; which must
remind us of that venerable measure of size with which
we are all familiar, "as big as a piece of chalk."   The
only rule founded in reason and justice is, sufficient mind
for the occasion.   I hardly need to suggest, that to dis-
tribute a large estate equitably and judiciously among a
considerable number of persons and institutions, must
require a stronger and wider exercise of mind than would
be needed for the disposal of a small estate among the
few legally entitled to a share in it.   And as much may
be said of a contract involving many contingencies, as
compared with one disposing of a few acres of land.

"Thus far, it will be seen, the elements of testamentary
capacity are strictly intellectual, pertaining to the pure
reason.   A testament, however, is not always the product
exclusively of the understanding.   The moral part of our
nature—the sentiments, affections and emotions—may be
as potent an agency in its production as the intellectual.
Hopes and fears, attachments, new and old, a sense of
dependence, a chronic habit of submission and deference,
assiduous attentions, crafty insinuations — these may
greatly prevail over the most obvious claims created by
the law of inheritance.   Inducements of this character
are not excluded by the law.   All that the law requires
in feeble-minded people is, that they be not excessive,
calculated to drive the mind from well-chosen, well-
matured arrangements, and divert the course of property
into channels it would not otherwise have taken.

"In examining these cases, we must never forget that
both the intellectual and the moral faculties may have
been concerned in their production, because, if either of
these factors are left out of the account, we are greatly

in danger of being misled. Nor must we forget, while investigating their respective agency, that they may act with some degree of independence of each other. The same person who thinks correctly and sensibly so far as he goes, may readily yield to inducements strongly presented to his feelings. In a case lately tried in Massachusetts, this distinction was so sharply and so pertinently made, that it may claim a moment's attention. A man, never married, confessedly feeble-minded, and under guardianship, concluded to make his will, which he did without urging or hindrance from any quarter. In this he bequeathed his property — $200,000 — to various charitable institutions, and to a few intimate friends from whom he had received much care and kindness, completely ignoring his sisters, for the reason, as it appeared in evidence, that they took no notice of him and were wealthy enough already. In making the will he was aided by a lawyer of the highest moral standing, and the whole transaction was free from suspicious circumstances. Two or three years afterwards, while in company of certain persons whose relations to him gave them much influence over him, he became so much excited by their reproaches and solicitations that he called for the will and drew his pen through his signature. The will was offered for probate ; but the judge declined to approve it, and an appeal was made to the Supreme Judicial Court, in which a trial ended in the establishment of the will. The jury virtually said the will was a rational act, rationally and calmly done, with memory and judgment sufficient for the purpose ; but the revocation was done in a moment of passion, excited by the suggestion of others and too strong for his feeble mind to resist. The verdict of the jury was approved by the court.

"The question of testamentary capacity in feeble-minded people is generally connected with that of outside influence. Sound-minded people may and often do ask advice in the final disposition of their property, and

the result is, very likely, all the better for it. Such
advice may be needed all the more where the mind is
weakened by nature or disease ; but when obtained, it is
always, and very justly, viewed by the law with suspicion, and the dominant question is, whether or not the
testator has been subjected to what is called *undue
influence* — because, sound or unsound, strong or weak,
his will must be his own will, and not another's.   If the
influence is such that the wishes and the interest of
other parties, rather than the testator's, are represented,
then the law supposes that the will is really not his will.
In order to establish the fact of undue influence, however, a foundation must be laid by first proving the fact
of mental deficiency.   The attempt is sometimes made
by lawyers, aided by physicians, to reverse the process.
Unless the testator has clearly manifested some mental
unsoundness by his previous acts, the proof of undue
influence should be strong enough to be unmistakable.
It often happens that no such can be shown ; that up to
the moment of the making of the will no indications of
feebleness, of delusion or wandering, have been witnessed.   I do not say that even under such circumstances undue influence may not be exerted.   When we
consider the enfeebling effect on body and mind of a
long last illness, of the many infirmities that often
attend this period, and of the utter prostration of the
will produced by pain and a sense of complete dependence, we can scarcely conceive of conditions better
fitted for the exercise of an undue influence over testamentary dispositions.   The courts of our time have
become quite familiar with a certain class of cases presenting these traits.   An old man marries a young
woman, and within a year or two dies, leaving a will
greatly in her favor, much to the disappointment of
relatives who would otherwise have received the whole
of the estate.   These cases are exceedingly embarrassing,
for we are often left without any clew to guide us to a
rightful conclusion.   We are sure it is such a will as the

testator would not have made in the vigor and flush of
health ; while we hesitate to say, under the conviction
that a man has a right to do what he pleases with his
own, how far a sense of gratitude for kindness and service may be allowed to shape his decision.   Fortunately,
perhaps, it frequently happens that some circumstance
sheds a little light upon the case, enabling the jury, if
not to decide according to its legal merits, yet to do
what, in their rude estimate, 'is about right.'

"You observed, no doubt, that Coke, in his classification of mental disorders, ignores entirely a form of the
disease which is far from being very rare.   The reason
was, probably, that it seemed so partial in its operation
—it left so much of the mind free, apparently, from its
influence — that it was not supposed to impair the person's responsibility for civil or criminal acts.   Nearly
a hundred years went by before it was first formally
recognized by Lord Hale, as well as another hundred
years after him before the law began to take it into the
account as an element of excuse for human delinquency,
or regard it as a claim on its protection.   I refer to what
is now called partial insanity, in which a person, while
entertaining some notions having no possible existence
in all other respects, talks and acts like other men.
How to meet the difficulties suggested by this form of
mental disease, was a problem entirely beyond the reach
of this luminary of the English law ; and he passed it
along to his successors, many of whom, even to this day,
have been as unwilling as he to give it any practical
effect in the administration of justice.   The light shed
upon the nature of disease, and especially of insanity,
by our better methods and opportunities of study, has
not been utterly disregarded ; and though we are too
often obliged to witness the display of the old ignorance,
rather than the new knowledge, yet the time is coming,
it is to be hoped, when the law will be in entire accordance with science.

"It is somewhat curious, that up to the present

century we hear nothing of partial insanity, strictly so
called, in civil cases ; and in criminal cases, where it was
sometimes offered in defense, it scarcely received a
respectful hearing.   The common idea that the monomaniac had still reason enough left for all practical
purposes, protected his testamentary privileges ; and
generally, it may be supposed, the rule worked no
injustice.   Had the courts been brought face to face with
a will the manifest offspring of a gross delusion, they
might have refused to sustain it.   It is hardly conceivable
that they would have approved a will devising a large
estate to the building of a railway to the moon (though
hardly more of a folly than many exclusively mundane
that have been built), however prudent, sagacious and
intelligent the testator may have been in the ordinary
affairs of life.   It was not until 1828 that this question
of the legal effect of partial insanity on the testamentary
capacity was squarely met and rightly decided.   I refer,
of course, to the case of Dew v. Clarke, which came up
for final adjudication in one of the English Ecclesiastical
Courts, Sir John Nicholl presiding.   The matter in issue
was the validity of the will of a London surgeon, who
bequeathed the bulk of his estate to his nephews, leaving
only a life interest in a small portion of it to a daughter,
his only child.   The testator had, for many years, very
creditably practised his profession ; and though regarded
by his patients and many others as eccentric and irritable,
was never suspected by them of laboring under any kind
of mental derangement.   On the other hand, it appeared
that he had always entertained the strongest aversion
towards this daughter, describing her to his friends and
strangers as prone to all manner of vice — as a perfect
fiend, an imp of Satan — charging her even with impossible crimes.   His treatment of her was almost incredibly
savage, not only compelling her to perform the most
menial offices, but he would often strike her with his
clenched fists, cut her flesh with a horse-whip, and once,
when she was only ten or eleven years old, he stripped her

naked, tied her to a bed-post, and after flogging her
with a large rod intertwisted with brass wire, rubbed her
back with brine. It appeared that there was no cause
for this extraordinary antipathy. The girl was described
by all who knew her as amiable and docile in her disposition, and perfectly correct in her deportment. Against
this will the judge pronounced an exhaustive and most
elaborate judgment, untrammelled as he was by the pre-
cedents of the common law, and inspired by the larger
spirit and freedom of the civil law. Remarkably well
informed on the nature of insanity, he discussed its effect
on the mind of the testator with a sagacity never before
witnessed in a court of law, reaching to the conclusion
that the mental disorder was fatal to the validity of the
will. Against the doctrine there announced, novel and
unprecedented as it was, no voice of dissent has ever
been raised. It is one of the few things in the medical
jurisprudence of insanity which may be considered as
established.

"Sir John Nicholl, be it observed, was careful to
restrict the operation of the principle to the case before
him. The mental disorder was sufficient to vitiate, not
any will, but the will in question. Had the testator
bequeathed his property to his daughter, he would
probably have established the will, insane as he was.
The insanity would have been no bar to a natural and
proper distribution of the estate ; and so, I apprehend,
the matter is now generally regarded. The objection
arises only when the distribution is not deemed to be
natural and proper by various relations, who find it for
their interest to destroy the will. The principle being
settled that insanity does not necessarily impair a man's
capacity to make a will, any more than it destoys his
power to do many other things as well as ever, its effect
ought not to be determined by any arbitrary rule, but
rather by that judicious consideration of the various circumstances of the case which is founded on correct views
of the nature of insanity and the ordinary motives of

human conduct.   In accordance with these views, a case
was adjudicated in the Court of Queen's Bench in 1870,
Banks v. Goodfellow, L. R. v. Q., 549, Chief Justice Cock-
burn delivering the opinion of the court.   Here a will
was established, notwithstanding the testator was proved
to have entertained some gross delusions, for it was
obvious that these delusions could not possibly have
influenced the dispositions of the will.   These two cases,
I presume, have settled the rule of law in regard to the
effect of delusions on testamentary competence, and
thus, happily, brought the law of the land into harmony
with the laws, physiological and pathological, of the
mental constitution.   And let me say in this connection,
that the effect of mental impairments on the testamentary
power is not to be estimated solely by their demonstrative symptoms, for it may be greatest when scarcely seen
by the world at large.   I have never met with worse wills
than some made under such circumstances.   I have
known a will made shortly after an apparently slight
attack of paralysis, pre-eminently absurd, irrelevant to
any worthy purpose, and almost if not quite impracticable of execution.   And yet the testator seemed to
have completely recovered, and continued his ordinary
pursuits till prostrated by a second attack.

   "Wise and proper as the doctrine may be, generally
stated that a delusion should vitiate any testamentary
disposition made under its influence, cases can easily be
imagined where it would be exceedingly embarrassing to
determine the exact range of its application.   We readily admit that the will of a man disinheriting all his
heirs-at-law—brothers, sisters, nephews and nieces—in
the belief that they have been attempting to take his
life, should not stand.   But supposing this delusion
referred only to a single relative, the rest of them being
properly remembered, I think we should hesitate to
break the will for that reason alone.

   "In pursuing the progress of thought on this subject,
we meet at last one of the extravagances of opinion,

which, coming from men of commanding intellect, produce surprise, if not admiration. Lord Brougham declared on one occasion that partial insanity, however limited apparently, as well as the more general forms of the disease, should vitiate all the patient's civil acts. He regards the mind as a single, indivisible potency, and consequently that any impairment of it must be absolute, not partial. On this theory, of course, there is no place for the practice of dividing and sub-dividing the mind, some portions becoming unsound, while others remain sound. Lord Brougham's doctrine is not without warrant, certainly, in the prevalent metaphysical theories of the last century, and, accepting them, it would be easier to reject it with feelings of wonder and surprise than to refute it. If inconsistency would furnish a conclusive argument against it, it may be found in the statement he once made, that a man might be so unsound as to be regarded by his Maker as irresponsible for criminal acts, while he might be justly held responsible by his fellow-men.

" And here we see the injustice that might be committed by making insanity, abstractly speaking, incompatible with testamentary capacity ; for if we say that a man who disinherits his heirs-at-law under the delusion that they have attempted to poison him, is thereby *non compos*, how shall we answer the question whether his will should be approved, even if he had bequeathed his property to those heirs-at-law, notwithstanding his delusions ?

"The effect on testamentary capacity of extraordinary beliefs, fanciful projects, or bequests for impracticable purposes, is frequently not very easily determined. Such things are suggestive of insanity, and the event has sometimes been made to turn on nice distinctions between insanity and eccentricity. In these cases the proper line of inquiry must depend on the circumstances of each particular case, and the decision should be governed more by the dictates of common sense, than any aribitrary rules of law. In some cases there can be little difficulty in arriving at a satisfactory conclusion. If a man noted for some oddities of thinking and acting, but otherwise correct and shrewd, believes that Brandreth's pills are a certain cure for all diseases whatever, and that everybody who would take enough of them would live to a good old age, this notion would hardly vitiate a will making unexepected and unjust bequests having no connection with and tracable in no way to it. If, on the contrary, he had devoted a considerable portion of his estate to the maintenance of a fund for supplying the poor with Brandreth's pills, this certainly would be a good reason for breaking the will.   Take another case. In Massachusetts, lately, an elderly gentleman in failing health, and with divers nervous ailments, was induced to try the movement cure, and came at last to conceive the most exaggerated notions of its medical efficacy, though it never helped him much.   Indeed, some of these notions almost, if not quite, amounted to delusion. In this state of mind he made his will, by which he appropriated a great part of his estate to the establishment of an asylum for nervous invalids to be treated by the movement method.   I have no hesitation in saying that that will was the offspring of a morbid nervous condition, if not of delusion, and therefore not to be established.   Whether certain mental manifestations are indicative of insanity or only eccentricity, is a point not always easily settled ; and no splitting of hairs on the question will prove so satisfactory as the exercise of a little common sense.   In many of these cases where, apparently, the mental twist is very limited and of doubtful character, a close scrutiny of the conduct and conversation will show here and there traces of a more extensive influence, thus shedding additional light on the matter in hand.

  "In presenting the subject of testamentary capacity in the way I have, it was for the purpose of giving to the pathological element the prominence it rightfully deserves, and which consequently ought to secure it a controlling influence in disputed cases. And let me say, in conclusion, that the administration of justice in this particular must often be imperfect, until the light of medical science is freely admitted and used — not the light that has traveled down to us from the times of Coke and Hale, but that which we owe to the progress of knowledge during the present century—greater, far greater, indeed, than that of all other centuries together.''

My honored friend, the Hon. George H. Yeaman, of New York, says respecting *spiritualism in wills:*

"The late case of Middleditch *v.* Williams, in the New Jersey Prerogative Court, decided by Van Vleet, Vice Ordinary, is of very great importance upon the subject of wills and their validity.

"The will was made by the testator under the belief that its provisions were desired and directed by the spirit of his deceased wife, communicated through a medium.

"The learned judge correctly holds the following propositions :

"1. That a will may be contrary to the principles of justice and humanity, yet if there was testamentary capacity, and no fraud or undue influence, the will must be upheld.

"2. That believing a thing upon false evidence, or insufficient evidence, is not an insane delusion.

"3. That an insane delusion is where a person conceives something extravagant to exist which has no existence, something which springs up spontaneously in the mind, and is not the result of evidence of any kind.

"4. That mania does not *per se* vitiate any transaction, for the question is, whether such transaction has been affected by it. (Of course monomania, sometimes called 'partial insanity,' is here meant; and thus the case only states the familiar doctrine that to vitiate a will the monomania must have influenced or determined the testator in making the will.)

"5. That belief in Spiritualism — that is, that the

11

spirits of the dead can and do communicate with the living — is not an insane delusion.

"The will was upheld. Every proposition as above stated is sound law. Yet it is submitted that the conclusion reached was incorrect, and that the judgment was the result of a failure to allow the 'elasticity' of the law to expand and adapt itself, not only to real advances in modern life, but to guard against things which, because not demonstrable, are at least not a part of life, so far as we know it; things which in the present state of knowledge cannot be made the basis of morality, social duties and legal rights. The system of legal reasoning adopted by the court was well established before modern Spiritualism had become a belief among men and women, and was adopted in order to reach a correct determination, in each case, of the question whether the paper offered was *the will* of the deceased. That is the ultimate question for solution in all cases.

" In the case referred to it is manifest that there was enough to invalidate the will, except for the ruling that a belief in Spiritualism is not *per se* insanity, not even monomania. And yet this belief gets upon record a will *that was not the will of the deceased.* That is not a satisfactory result.

"The case can perhaps be better understood by a few quotations :

" 'The testator was a believer in Spiritualism ; that is, he believed the spirits of the dead can communicate with the living through the agency of persons called 'mediums,' and who possess qualities or gifts not possessed by mankind in general. The proofs show that the testator stated to several persons, prior to the execution of his will, that the spirit of his dead wife had requested him, through a medium residing in Forty-sixth street, in the city of New York, to make provision for his mother-in-law in his will. To one person he said that his wife's spirit had requested him to give all his property to her mother, and to do it in such a way that none of his relatives could get it away from her. To the same person he said, at another time, that the spirit of his wife was constantly urging him to make a will in favor of her mother. To another person he said that the spirit of his wife had requested him to be good to her mother, and see that she was made comfortable during the remainder of her life, and he also said he intended to make a will, leaving enough to his mother-in-law to make her comfortable, because his wife wanted him to do so.'

*   *   *   *   *   *   *   *

"'The utmost length to which any court has as yet gone on this subject is to declare that a belief in Spiritualism may justify the setting aside of a will, when it is shown that the testator, through fear, dread or reverence of the spirit with which he believed himself to be in communication, allowed his will and judgment to be overpowered, and in disposing of his property followed implicitly the directions which he believed the spirit gave him ; but in such case the will is set aside, not on the ground of insanity, but of undue influence. Thompson v. Hawks, 14 Fed. Rep., 902.'

*   *   *   *   *   *   *   *

"'There is no evidence in this case which will support a conclusion that the testator, at the time he executed his will, was subject to an insane delusion. Nor do I think there is any evidence in the case which will support a judgment declaring that the will in question is the result of undue influence. There is no proof tending to show what influence the spirits or medium exercised over the testator in making his will, except that which proceeded from the testator's own mouth. His declarations are competent to show the condition of his mind, but not to prove undue influence against either persons or spirits. Rusling v. Rusling, 36 N. J. Eq., 603-607.'

*   *   *   *   *   *   *   *

'Neither the medium, nor Mrs. Williams (the mother-

in-law), nor any other person who was present at any
of the *seances*, has been examined as a witness. No
legal evidence of what occurred at any of them is before
the court. The charge of undue influence is mainly
directed against Mrs. Williams. She is said to be a
believer in Spiritualism, and the proofs show that she
went with the testator frequently when he went to the
medium to consult the spirit of his dead wife. There are
some things in her conduct which are calculated to create
strong suspicion. Without apparent cause she seems to
have entertained feelings of strong dislike toward all the
testator's relatives.'

*   *   *   *   *   *   *   *

"These things naturally breed suspicions and create
fears. They show that it is possible that every message that the testator received, purporting to come
from the spirit of his dead wife, came not from the
dead but from the living; and that every thing that was
done to dispel the testator's doubts, and to induce him to
believe in the reality of the spiritual manifestations
which he witnessed was, from beginning to end, a pre-
arranged scheme of deception and fraud. But there is
no proof in the case which will support a judgment that
such was the fact. There is enough to raise a strong
suspicion, but not enough to produce conviction. Undue
influence, like fraud, cannot, in a case where no relation
of trust exists, be presumed, but must be proved. I
strongly suspect that the testator was duped. It may
also be true that he was unduly influenced. I believe
that the examination of Mrs. Williams or the medium,
as a witness, would in all probability have made many
things which now seem dark and obscure plain and clear.
The question, however, whether or not the paper in question is the will of the testator must be decided by the
evidence before the court.'

"The importance of this judgment cannot well be over-
rated. The rule it adopts, if approved and followed, will
add a new stimulus, increased activity, ingenuity, avarice

and rapacity to the army of 'spirits' that talk or write to us through 'mediums.' The family of this testator was robbed and plundered by them. How many other families are yet to be victimized in the same way?

"Law, in seeking to ascertain facts, for the purpose of basing judgments on those facts, must keep within the domain of verifiable human knowledge. Here the leading fact, well enough established, is that the testator believed in Spiritualism, *and made his will under the influence of what he supposed a communication from the spirit of his dead wife.* The mildest thing that can now be said of Spiritualism is that it is not known to be true, and probably cannot be proved to be true. Therefore a will that comes into court out of the spiritual world does not come with the verifiable safeguards required by the law. But suppose that Spiritualism be not only true, but that it has been thoroughly and indisputably proved to be true—has become as much an accepted fact as the Copernican system of astronomy, or the circulation of the blood, two discoveries for which great leaders in science suffered—then what? Can it be allowed that a spirit, over whom we have no more control than we have over the winds, shall send a will into court, call it my will, and have it probated as such, give my estate to strangers and beggar my children? Law is of the earth, earthy. It is for things as we see them, to control things as we know them. If our estates are to be disposed of by 'spirits' speaking through 'mediums,' why shall we not open a spiritual polling place and go there and vote as directed? And if our departed friend was entitled to vote while visible here on earth, incumbered by that despicable tabernacle of clay, why not open the ballot-box to the reception of his gostly ballot on election day, now that he has become so etherealized and sublimated as to rise superior to all the base considerations that might have influenced him here? True, this would require an amendment of the Constitution and statutes. That is to say, spirit voting is not now provided for and allowed by law. No more is spirit *will-making*—at least, not that spirits may make wills for other people. Let them devise their own property, if they have any, and send their wills in for probate when they die— though their dying might shake the faith of some who now consult them.

"Could not the learned judge have reached a correct decision by a short cut? Thus: Belief in Spiritualism is not insanity, because it is merely believing on false or insufficient evidence. But if a man makes a will under such belief, then his will is made under an "undue influence," because it is an influence which courts, in our present state of knowledge, cannot recognize except in its results which are the same as of insanity. Or thus: Belief in Spiritualism is not insanity; but a will made under the belief that a spirit directed it is nevertheless made under the influence of monomania, because while believing in the fact of spiritual communication is not insanity, yet believing in the right of a departed spirit to rob a man's children, to enrich his mother-in-law, is an insane delusion. Or thus: 'Belief in Spiritualism is not insanity. Indeed, Spiritualism is an established fact. But we (the court) are not yet in the spirit land. We have to deal with men and things as we know them and as the law directs. The question is, was this will freely made by a *man* who once lived but is now dead? His spirit has not come back to tell us that he really made the will; and it is in proof before us that he thought the spirit of his dead wife told him to make it as he did. We most implicitly believe that she did so. Nay, we will go further and say that we believe the law ought to provide for legalizing such wills. But the law is conservative. It moves slowly. It lags behind the advanced thought, and even the necessities of an improving civilization, an ever-widening intelligence ; and as we are here to administer, and not to make the law, we reluctantly refuse the probate of the paper in question as the will of the deceased.'

"If either of the three forms of reasoning would be new, so are the facts, and facts are the things we have to deal with in this unspiritual world.

"The case under review excites reflections on problems that have been pondered by men since the dawn of history — how much longer we know not. Some one has said: 'Existence is the sum of mysteries.' And so it is. If we try to account for life by 'evolution' out of dead matter, the matter yet needs to be accounted for. If we account for both by reference to an Almighty intelligence, power and will, the mystery remains; for experience and holy writ unite in testifying that 'no man hath seen God.' Some one else has said that 'we know more of spirit than we do of matter.' Assuming the truth of the duality of existence implied in the proposition, it is true, at least as to things in this life. We know absolutely nothing of the ultimate nature and constitution of the penholder in our fingers, or the ink and paper we spoil in trying to express our dissent from a learned judge's decision. But we know that we doubt. And herein lies our greater knowledge of spirit than of matter. Des Cartes said: '*Dubito, ergo Cogito.*' But he said a far greater and more momentous thing: '*Cogito ergo sum;*' three words that weigh more than some entire volumes on 'metaphysics.'

"'I am' and 'I think' are the best known facts of existence. Given the truth of duality, the separateness of mind and matter, in duality and in kind, we at once perceive that it is just as possible that the individuality, the personality, expressed in 'I think,' may exist after the death of the body as that it should exist now. Most people believe that it will. This belief is not ground for the imputation of weakness, much less of insanity. On the contrary, the common law held that a man who believes and also believes in future punishment, is at once a safer witness for the truth and a better man than one who does not so believe. From this belief to a belief that departed spirits commune with us while we live here, is but a short step, if only there be any evidence whatever to sustain it. In the popular conception of it, 'evidence' is an extremely indefinite thing. What would convince some very good people would only cause a sneer among scientific men and trained lawyers. Nevertheless, it has convinced some, and hence this believing a thing on wholly insufficient evidence is properly held to be not an insane delusion; though it must be admitted that the fact that the spirit generally speaks, or writes, or raps through a 'medium' who is a tallow-faced hag, or a long-haired man with clammy hands, instead of communing directly with the loved ones here; and the fact that when the spirit does 'materialize' and move and talk *in propria persona*, a firm hug and a ruthless tearing away of drapery always reveals a woman still 'in the flesh,' and as much opposed to dying as the rest of us, ought to make the believers more sceptical — at least careful.

"If such a belief must be held to be no evidence of insanity, is there no mode of legal reasoning by which courts, without legislative intervention, can protect children and grandchildren against a mother-in-law to whom the spirit of the dead wife has implored the husband to give his property? Cannot such a disgusting fact be used as a side light while looking into the will itself for evidences of testamentary capacity or incapacity? The real question is, not whether Spiritualism is true, or whether believing in it is an insane delusion. It is whether a will made under its influence is a real will, or a safe thing for the world as we know it. Would there be any judicial violence done by courts saying plumply that when a man surrenders himself to this sort of unproved and unknowable thing and his will is traceable to the influence of this belief, that then he lacks testamentary capacity? Instead of standing in fear, dread and awe of the spirit, he loved its converse. And just because he loved it and was controlled by it, was it not an 'undue influence?' Are we to be put to the hazard of swearing the medium and the mother-in-law to show what occurred at the *seances?* And when, as of course, they testify that *they* used no influence, but only communicated a real message from a real spirit, are we to believe it? Is it not enough that the testator himself says he had the message? And suppose we do believe it, *and suppose it is all true*, is the case any better? Is Spiritualism, in any view of it, a part of the practical realities of this life, as we know them : a basis for ethics, for family settlements, for testamentary disposition? Does not the question remain as heretofore : *was it his will?* If the testator says he was thus influenced, what more do we want? If he was mistaken in thinking his will was dictated by the spirit of his dead wife, not to say *defrauded* into so believing, ought it not to be rejected? And *if in fact dictated by her*, ought it not equally to be rejected? It was not his will.''

Evidence of the Insane. Lunacy was, until a recent date, regarded by the law as incapacitating a patient from giving evidence in court. But according to the much more extended significations which the term lunacy has received, it now includes states of mind which are compatible with testimonial capacity. Where the judge is satisfied that the lunatic understands the obligation of an oath, and can give a rational account of such things as happened before his eyes, the evidence may be admitted. But the weight to be attached to such evidence will still depend on the extent to which it fulfils the conditions commonly required to constitute credibility. It has been held, however, that when a person has suffered from an attack of insanity between the occurrence of a transaction and the time he renders his testimony, his evidence cannot be admitted.

Opinions of Ordinary Witnesses as to Insanity. The statement of a non-professional witness, as to the sanity or insanity, at a particular time, of an individual whose appearance, manner, habits and conduct came under his personal observation, is not the expression of mere opinion. In form, it is opinion, because it expresses an inference or conclusion based upon observation of the appearance, manner and motions of another person, of which a correct idea cannot well be communicated in words to others, without embodying, more or less, the impression or judgment of the witness. But, in a substantial sense, and for every purpose essential to a safe conclusion, the mental condition of an individual, as sane or insane, is a fact, and the expressed opinion of one who has had adequate opportunities to observe his conduct and appearance is but the statement of a fact; not, indeed, a fact established by direct and positive proof, because in most if not all cases it is impossible to determine, with absolute certainty, the precise mental condition of another; yet, being founded on actual observation, and being consistent with common experience and the ordinary manifestations of the condition of the mind, it is knowledge, so far as the human intellect can acquire knowledge upon such subjects. Insanity "is a disease of the mind which assumes as many and various forms as there are shades of difference in the human character." It is, as has been well said, "a condition which impresses itself as an aggregate on the observer," and the opinion of one, personally cognizant of the minute circumstances making up that aggregate, and which are detailed in connection with such opinion, is, in its essence, only fact "at short-hand" (1 Wharton & Stille's Med. Juris., § 257). This species of evidence should be admitted, not only because of its intrinsic value, when the result of observation by persons of intelligence, but from necessity. We say from necessity, because a jury or court, having had no opportunity of personal observation, would otherwise be deprived of the knowledge which others possess, but also because, if the witness may be permitted to state— as undoubtedly he would be, where his opportunities of observation have been adequate—"that he has known the individual for many years; has repeatedly conversed with him and heard others converse with him; that the witness had noticed that in these conversations he was incoherent and silly; that in his habits he was occasionally highly pleased and greatly vexed without a cause ; and that, in his conduct, he was wild, irrational, extravagant and crazy—what would this be but to declare the judgment or opinion of the witness of what is incoherent or foolish in conversation, what reasonable cause of pleasure or resentment, and what the indicia of sound or disordered intellect? If he may not so testify, but must give the supposed silly and incoherent language, state the degrees and all the accompanying circumstances of highly excited emotion, and specifically set forth the freaks or acts regarded as irrational, and thus, without the least intimation of any opinion which he has formed of their character, where are such witnesses to be found ? Can it be supposed that those, not having a special interest in the subject, shall have so charged their memories with these matters, as distinct, independent facts, as to be able to present them in their entirety and simplicity to the jury ? Or, if such a witness be found, can he conceal from the jury the impression which has been made upon his mind ? and when this is collected, can it be doubted but that his judgment has been influenced by many, very many circumstances which he has not communicated, which he cannot communicate, and of which he himself is not aware ?" (Clary v. Clary, 2 Iredell's Law, 83). The jury, being informed as to the witness' opportunities to know all of the circumstances, and of the reasons upon which he rests his statement as to the ultimate general fact of sanity or insanity, are able to test the accuracy or soundness of the opinion expressed, and thus, by using the ordinary means for the ascertainment of truth, reach the ends of substantial justice.

These views are sustained by a large number of adjudications in the courts of this country, some of which are here cited : Clary v. Clary, 2 Iredell's Law, 83 ; Dunham's Appeal, 22 Conn., 193 ; Grant v. Thompson, 4 id., 203 ; Hardy v. Merrill, 56 N. H., 227, substantially overruling Boardman v. Boardman, 47 N. H., 12, State v. Pike, 49

id., 468, and State v. Archer, 54 N. H., 498; Hathaway's
Adm'r v. National Life Ins. Co., 48 Vt., 350; Morse v.
Crawford, 17 id., 499 ; State v. Clark, 12 Ohio, 483 ; Gibson v. Gilman, 9 Yerg., 330 ; Potts v. House, 6 Geo., 324 ;
Vanauken's Case, 2 Stock. Chy., 190 ; Brooke v. Towns-
end, 7 Gill., 10 ; De Witt v. Barly, 17 N. Y., 342, explain-
ing decision in same case in 5 Selden, 371 ; Hewlett v.
Wood, 55 id., 634 ; Clapp v. Fullerton, 34 id., 190 ; Ruth-
erford v. Morris, 77 Ill., 397 ; Duffield v. Morris, 2 Har-
rington, 384 ; Wilkinson v. Pearson, 23 Pa. St., 119 ;
Ridcock v. Potter, 68 id., 342 : Doe v. Reagan, 5 Blackf.,
218 ; Dove v. State, Heisk., 348 ; Butler v. St. Louis Life
Ins. Co., 45 Iowa, 93 ; People v. Sanford, 43 Cal., 29 ;
State v. Kringer, 46 Mo., 229 ; Holcombe v. State, 41
Tex., 125 ; McClackey v. State, 5 App. (Tex.), 320 ; Norton v. Moore, 3 Head., 482 ; Powell v. State, 25 Ala.,
23 ; 1 Bishop's Crim. Pro., secs. 536-40 ; 1 Wharton &
& Stille's Med. Juris., sec. 257 ; Wharton's Law of Evidence, secs. 510 *et seq.;* 1 Redfield on Wills, ch. 4, part 2,
in a recent edition of which (p. 145, n. 24) it is said, touching the decision in Hardy v. Merrill, *ubi supra:* "There
will now remain scarcely any dissentients among the elder
States ; and those of recent origin, whose decisions have
been based upon the authority of the earlier decisions of
some of the older States which have since abandoned the
ground, may also be expected to change." See also
May v. Bradlee, 127 Mass., 414 ; Com. v. Sturtevant, 117
id., 122. In several of those cited the whole subject was
very fully considered in all its aspects. While the cases
are, to some extent, in conflict, we are satisfied that the
rule most consistent with sound reason, and sustained by
authority, is that indicated in this opinion.

As to the admissibility of the opinion of witnesses in
matters of testamentary capacity, we would refer to the
best opinion on record, that of Judge Doe, in State v.
Pike, 11 Am. Law Reg. (N. S.), 223, 241 ; 51 N. H., 105,
quoted at length on p. 115, Jarman on Wills. Witnesses
not experts should be allowed, we think, to testify that,

from their observations of the testator's appearance and conduct, they formed the opinion that he was either sane or insane. It is, we think, clearly competent evidence.

Management of Property. Where persons are supposed to be unable, from unsoundness of mind, to undertake the management of their own property, it may be necessary that they should be placed under the protection of the court; but this proceeding is not usually had recourse to, unless there is urgent necessity, or there is a strong probability that the person's incapacity will be permanent. It is resorted to principally in chronic or congenital cases, where there is no room for doubt as to the mental condition of the individual; and in cases of recent insanity, where it is necessary to have recourse to an asylum for the protection of the individual, it may also be necessary to obtain protection for his property by the aid of the court. In giving evidence or framing a statement in such a case, it is important, if incapacity is to be proved, to show that the individual has been found, when placed in circumstances requiring such capacity, unable to perform the acts which the management of property necessitates. In cases of active insanity, it is especially required to show not merely that there is delusion or other symptom of insanity, but that the insanity is of such a nature as specially to disable the person from duly performing the duties which would be required of him. Difficulties most frequently occur in cases of imbecility and dementia; but the verdicts in such cases, when disputed, will generally be found to rest rather upon the impression produced by evidence of the actual behavior of the individual than the mere medical view of his mental condition. The most effectual aid that his medical witness can render in such case, is to show whether there are or are not such peculiarities in the conduct of the person under inquisition, as are known to be characteristic of imbeciles or demented persons. In undisputed cases, where the duty of the physician consists merely in making an affidavit, there is special difficulty to be encountered. Brevity, scrupulous accuracy, and attention to the fact that such unsoundness of mind as involves incompetency to manage property must be established, are the most important requirements. In England, a person found by the court to be incapable, is placed under the control of a "committee of the person," and the property under a "committee of the estate." In Scotland, an application to the Court of Sessions for the appointment of a *curator bonis*, takes the place of the English inquisition. The chief peculiarities of the Scotch process are, that it is cheaper, more easily effected and more easily annulled, and that it does not effect the person of the lunatic. By the provisions of a recent act, the person of an insane man in Scotland may be placed under the guardianship of the nearest male relation found competent.

Legal Tests of Responsibility. Bucknill and Tuke say in respect to this, that although in practice the plea of insanity in criminal cases is in a large number of instances not determined according to the law laid down by judges, but according to the higher law of humanity, that it is important that students of psychological medicine should know what unfortunately continues to be the main legal test of responsibility in criminal cases—the consciousness or knowledge of right or wrong ; instead of being as it should be, whether in consequence of congenital defect or acquired disease, the power of self-control is absent altogether, or is so far wanting as to render the individual irresponsible. As has again and again been shown, the unconsciousness of right and wrong is one thing, and the powerlessness through cerebral defect or disease to do right is another thing. To confound them in an asylum would simply have the effect of transferring a considerable number of the inmates thence to the treadmill or the gallows.

For cases in which the prisoner was acquitted on the ground of insanity, although knowing the nature and quality of the act and quite conscious of the difference

between right and wrong, the reader is referred to
Taylor's "Medical Jurisprudence," 4th ed., p. 768. For
cases in which the plea of irresistible impulse was admit-
ted, see p. 760, also p. 262-3 of Bucknill and Tuke's
"Manual of Psychological Medicine," 4th ed. Also
refer to "the case of Henry Galbites," by Dr. Kitching
("Journal of Mental Science," July, 1867); the same
writer's lecture on moral insanity ("British Medical
Journal," 1857); "The Legal Doctrine of Responsibility
in Relation to Insanity," by S. W. North, M. R. C. S.
("Transactions of the Social Science Association, 1864");
"Insanity and Crime," by the editor of the "Journal of
Mental Science," 1864 (Townley's case); "Étude Médico-
légale sur la Folie," par M. Tardieu, 1872; the work of
Esquirol and Marc, Brierre de Boismont, De la Folie,
Raisonnante, etc., 1867; "De la Monomanie de Persecution au point de vue de la Médecine légale," (Ann.
d'Hyg., pub. 1852), and Lasègue, "Memoire sur la
Délire des Persecution" (Arch. Gen. de. Méd., tom. 27).
A case of delusion of persecution ending in homicide and
acquittal, in which the judge's common sense and
humanity got the better of his law, will be found in the
"Journal of Mental Science," for July, 1872. For cases
proving the presence of the homicidal *impulse* without
other symptoms of insanity, see article by Dr. Needham
in the same number; and for the important cases of
Edmunds and Watson, see April, 1872. For case of
insane infanticide and the judge's summing up, see
April, 1871. Mr. J. B. Thompson's article in the
"Journal," January, 1870, and also the succeeding one
in the October number, which should be read in connection with Despine's work, "Psychologie Naturelle,"
1868.

Hints in Giving Evidence. Bucknill and Tuke say
respecting this part of medical jurisprudence:

"1. That a medical man is obliged to make known, if
asked in court, the statements or confessions made by a

patient to him (Peake on Evidence, p. 88 ; Starkie on Evidence, p. 105; Shelford, p. 81).

"2. If a medical witness believes a criminal to be insane and is called upon to give evidence to that effect, he must not be content with stating his opinion, but must be prepared to state the reasons upon which that conclusion is based. For aid in arriving at a judgment the reader is referred to the chapter on the Diagnosis of Insanity.

"3. The medical witness should confine himself to a simple statement of facts, and not allow himself to be drawn into a metaphysical discussion, or an attempt to define insanity.

"4. If a medical witness sometimes wishes to fortify his view of the case by inducing the counsel to read from medical works, and the question arises whether this can be legally done. It has been decided in one case that 'counsel was at liberty to read, as part of his speech, the opinions of a medical work, but the jury would not have to decide the case upon medical criticism, but upon the case and the facts.' The counsel in the case alluded to then read from a book on medical jurisprudence, in order to show that certain cases recorded there were similar to the one before the court. It would appear, from R. v. Crouch, 1 Cox, C. C., 94, that the *opinions* of a medical writer cannot be stated in an address to the jury, but the judge in the case alluded to did not distinguish between these and *cases*.

"5. In regard to any notes the medical witness may have taken of the prisoner's state, he may only make use of those in court which he has committed to paper at the time he examined the prisoner.

"6. It must not be forgotten that the prisoner may be sane when examined by the physician, and yet may have been insane when he committed the deed, and *vice versa.*" *

---

* See case of The People against Mrs. McClusky, who killed her child by throwing it from a window during delirium tremens; tried

Expert Testimony and the Functions of Experts.
Many of the community, as the late Dr. Ray has shown,
completely ignore the exact purpose of skilled testimony in a judicial proceeding and the functions of an
expert. They are apt to bring forward the timeworn
objection to expert testimony, viz., that as the experts
are engaged by one or the other of the litigant parties,
they thus necessarily testify under a bias, and consequently are not trustworthy. This would imply that
there is a distinct understanding as to what any given
expert shall say, before he has heard a word of the evidence on either side. An expert's opinions, as Dr. Ray
has said, are worth money, but it does not follow that
his opinions are corruptly bought. Why should a fair
reward for professional services obscure an expert's perception of truth ? Experts necessarily, according to the
present law, which we hope to see reformed (see our
writings on "the necessity for a reform in the introduction of expert testimony where insanity is alleged as a
defense"), testify in the interest or a party; but that
fact Dr. Ray conclusively proved does not imply an
unworthy bias. The counsel lay before the expert the
evidence to be produced before him as far as they can,
and the honest expert invariably tells the counsel either
that if he can prove the facts as he states them he has a
good case, or he tells him that even if he does prove
such facts they would not warrant the construction he
wishes to put upon them, and that his — the expert's —
testimony would not help him. Generally speaking, it
is as Dr. Ray said, that if an expert's testimony is
wholly and unconditionally in favor of one side only, it
is merely because this result is warranted by the facts.
An honest expert will, moreover, warn the counsel that

---

before Judge Gildersleeve, General Sessions, June 10, 1888. The jury
in this case, without retiring, promptly acquitted the prisoner as "Not
guilty, by reason of insanity." She had been a case of dipsomania.
of previously good character, and a kind and affectionate mother.
The writer appeared as the medico-legal expert for the defense.

the evidence as brought out on trial may oblige him to modify his opinion.

An expert is one who gives his time and attention entirely to a particular pursuit, and he is, therefore, to be recognized as an expert in questions relating to that pursuit, to the exclusion of those who have attended to it incidentally as a subordinate part of a more general department of inquiry.

The functions of an expert are to appear in court to give an opinion, based either on his acquaintance with the party whose mental or physical condition is under investigation, or upon a medical examination of him which he has made, or upon a hypothetical case stated to him in court. The expert is wanted in court to give his opinion on facts proved or upon a case hypothetically stated. *An opinion*, I should define as *the statement of what certain facts indicate to the expert himself*. Therefore, on a trial, I do not think an expert should give his opinion upon facts proved by a witness unless he hears all the testimony of such witness. The old practice where the expert heard all the evidence given at the trial, and then was asked for his opinion founded on that evidence, supposing it to be true, was, I think, better calculated to elicit a well-considered opinion than the new change, where the counsel on each side, out of the facts that have appeared in evidence, construct a hypothetical case as fairly as will best serve their purpose, and no more. In such cases the expert may be obliged to assent to the propositions of both sides, as Dr. Ray has shown, and thus apparently stultify himself. This is due to a twisting and coloring of facts. Sometimes, unfortunately, the manner in which an expert's opinion is elicited is deliberately calculated to overwhelm it with discredit. Able counsel use all their professional astuteness to deprive of its proper weight with the jury, says Dr. Ray, the most honest and truthful expressions of opinion, and if we had a healthier public sentiment which would make the judge keep a cross-examination within its proper limits and restrain the license of counsel, the public would have less reason for distrusting and sneering at expert testimony. The judge's question to determine whether a witness offered as a mental expert has the legal qualification to entitle him to testify as such should be : "Do you give your time and attention entirely to a particular branch of medicine, and is that mental or psychological medicine?" This, and nothing else, is needed to constitute an expert in mental medicine. He should then, in any given case, give his opinion on the case from the examination he has made ; his observation, experience and professional reading. He necessarily forms an opinion from this combination.

Dr. Ray recommended, in 1873, that the testimony of experts be given in writing and read to the jury without oral examination. It would thus, he said, be deliberately prepared, its explanations well considered, and its full force and bearings clearly discerned. It would go to the jury on its own merits, no advantage being gained by either party by the superior adroitness of counsel in embarrassing the witness and pushing his statements to a false or ridiculous conclusion. It would work no injustice to either party, and it could be managed without inconvenience. There could be no difficulty in civil cases where both parties consent to such an arrangement. Dr. Ray says: "Judges should not, as they sometimes have been known to do, disregard their proper functions and assume the part of an expert, and, in cases of disputed sanity, pronounce a man to be sane and safe to be at large in spite of the declarations to the contrary of men long conversant with the discourse, conduct, ways, and manners of the insane." The whole subject of expert testimony needs to be lifted up to a higher plane than it now occupies, by the mutual efforts of lawyers, physicians, and public sentiment. The revolution in the management of the insane has produced among its legitimate effects a better knowledge of in-

sanity. Respecting written testimony I would add that
in the celebrated Parish will case, reported by Ray in
his " Contributions to Mental Pathology," p. ?16,
where Henry Parish, a prosperous New York merchant,
made his will in 1842, being then fifty-four years old, on
trial the surrogate wisely determined that the opinions
should be given in writing, with the understanding that,
though not clothed with the authority of legal evidence,
they would be carefully considered and credited with all
the weight to which they were really entitled. This
enabled the expert to utter, as Dr. Ray showed, what is
impossible in the usual method of examination and cross-
examination, his opinions and the reasons for opinions,
with that coherence and logical relation absolutely necessary to show their full force and significance. Mr.
Parish made his will in 1842; disposing of some $750,000.
He went to Europe in 1843, and had an apoplectic
attack, from which he shortly recovered and continued
as well, apparently as ever, both in body and mind,
until the 19th of July 1849, when he had another apoplectic attack, much more severe. In about a fortnight
he was out of immediate danger, but never recovered his
ordinary condition. His right side, including the upper
and lower limbs, was found to be somewhat paralyzed ;
the power of articulation was lost ; and his natural elasticity and vigor was gone. These traits continued with
little change until he died, in 1856. Epileptic fits
occurred within a few months of the apoplectic attack
of July, 1849, at intervals ranging from eight days to
six months or more. On the 29th of August, 1849, he
subscribed his cross in lieu of a signature to a codicil to
his will. On the 15th of September, 1853, a second codicil was subscribed in like manner, and on the 15th of
June, 1854, a third codicil, substituting his wife, in place
of Daniel and James Parish, as residuary legatee. These
codicils were contested in the Surrogate's Court on the
ground that when they were made the testator had not a
testamentary capacity. His mental condition during the

period between the attacks in 1849 and his death in
1856 was that of but a small measure of mental capacity.
He was reduced to an almost vegetative existence.   There
was ample proofs of mental infirmity, of dementia or
imbecility.   He was plainly an insane man, without
mind enough left to constitute testamentary capacity.

THE PSYCHOLOGICAL ASPECT OF THE ANDERSON WILL
CASE.—The last will and testament of John Anderson, the
millionaire tobacconist, was executed October 25th, 1879;
a codicil was added to this will on September 29th, 1881,
revoking some provisions of the will.   John Anderson
died on the 22nd of November. 1881.

The case was commenced about May 27th, 1882, and
was brought by Mary Maud Watson, a granddaughter of
John Anderson, her mother's name having been before
marriage Mary Louise Anderson, afterwards Mrs. Carr.
The action was brought to recover one undivided fifth
part of the property mentioned in the complaint, it being
part of the realty belonging to the estate of John Anderson, deceased.

The plaintiff claimed that the alleged will and codicil
were not, nor was either of them, duly executed ; that at
the time of the execution thereof respectively, the said
John Anderson, deceased, was not of sound mind or
memory nor capable of making a will or codicil thereto,
and that the execution of said alleged last will and
codicil, respectively, was procured by the undue influence, duress and restraint exercised upon him.

The only heirs-at-law and next of kin of the said John
Anderson, deceased, are and were John Charles Ander-
son, the only surviving son of said decedent : Kate
Anderson, of the city of New York, his widow : Laura
V. Appleton, wife of Edward J. Appleton, of Brooklyn,
N. Y., the only surviving daughter of said decedent ;
Fannie A. Barnard, Mary A. Wagstaff, wife of Alfred
Wagstaff ; Alice Barnard, George G. Barnard and John
Charles Barnard, the surviving children of Fannie A.

Barnard, deceased, a daughter of said decedent—said
Alice, George G. and John C. Barnard being minors, and
said Alfred Wagstaff being their general guardian;
Agnes Bryant and Amanda Bryant, the only surviving
children of Amanda Bryant, deceased, a daughter of
said decedent, and lastly, Mary Maud Watson, the plaintiff in this suit.   The testimony of the appellants discloses many indications of mental disorder, the most
prominent being as follows, viz: The case as developed
on the trial claimed to show on the plaintiff's side that
John Anderson, deceased, was not of sound mind and
memory at the time of making his will, for the following
reasons, viz., Mr. Anderson was a man between 50 and
60 years of age, who, by his own efforts, amassed a large
fortune.   He was a man of limited education, and on
reaching the age of 50 or 60, he was under the impression,
or stated that he began to be visited by ghosts or spirits.
That he supposed his boy, Willie, then dead, appeared
to him, and that he held communication with him from
time to time, or with his spirit after death.   That on one
occasion he handed to a man, who had saved this boy's
life, $100, saying that the boy, Willie, then dead, had
appeared to him, and had asked him to make that gift
of that $100.   That the decedent supposed that he was
haunted by the spirit of Mary Rogers, a girl formerly in
his employ, and who had disappeared, and that her spirit
gave him much trouble for some period of time, until
finally he announced that it was all right with her and
with her spirit, and that she gave him no more trouble.
He adhered to these news and would not be persuaded
that they were all delusions or imaginations.   That he
believed a certain investment would pay 25 per cent., because the spirit of Mary Rogers said so.   That he
believed the mother of the plaintiff illigitimate, and that
his wife was a prostitute or had been.   That, although a
man of large wealth, he lived with his wife and one
servant in a large house, which was but partly furnished.
That the house had steel shutters, which were closely

drawn at night. That he was afraid his food would be poisoned, and gave directions to keep the ice box containing food locked. That he gave directions that no brass pins should be around the house, because he was afraid of being poisoned or affected by the pins. That he gave directions never to unbolt the door unless it was first learned who sought admission, because he was afraid somebody would shoot him. That he believed there was a conspiracy on the part of his family, or some of them, to stab him or kill him, or both. That he believed his son was a thief and robbed him of a large amount of property in a house of prostitution, though his son was the residuary legatee under his will and inherited the bulk of his property. That he had exaggerated ideas of his ability to re-fashion the governments, or some of them, in Europe, or to fashion these governments into a republic. That he stated that he expected to be a man of much importance in such a republic. That he desired to go away from the world and be alone. That he was troubled with loss of sleep and severe pains in the head. That he had an intention to kill himself on that account. That shortly after making his codicil he left for Europe, with the remark that when he got away from these people he intended to make a different will. That late in life his walk was irregular; that his feet and hands shook, and left foot dropped and he walked with a halting gate. That he became weak and feeble in body, incoherent in conversation, passing from one subject to another, without apparent causes, from business to the discussion of the situation of spirits. That he made untrue and disgusting remarks about men or acquaintances. That he threatened or offered to expose his person in a public place in order to convince his friend that his physical powers were not impaired. That he had a shot gun, rifle and sword ready to do deadly injury to a son-in-law of his. That he died within about two years from the date of his will, and within a few months from the date of his codicil. That out of a fortune of several millions, he left the child of his daughter the income of $20,000. That at one time, late at 'night, he left the residence of his son-in-law, Geo. C. Barnard, in cold weather, in his stocking feet, without shoes and without a coat. and was afterwards, in that same night, discovered at the Astor House in that condition. That he refused to go to the residence of Judge Barnard because they had a conspiracy to kill him. That he tried to persuade a friend or a gentleman to go to Judge Hackett and converse with him on that account.

On the contrary, the witnesses to the will testified that in their opinion (?) said decedent was of sound mind and memory when he made the will.

The facts we relate were testified to by several witnesses, one of whom, Mr. McCloskey, had been acquainted with the decedent for thirty-five years and had been connected with him in business.

The great medico-legal point in this case is this : Did the delusions of the decedent (Anderson) influence the disposition of the will ? If so, the mental disorder was sufficient to vitiate the will in question. There, it seems to me, we are brought face to face with a will, the manifest offspring of a gross delusion.

A person to make a valid will, must understand perfectly the nature and amount of the property they are disposing of ; must have a sound disposing mind and memory ; must not ignore the natural claims of relationship and affection, and must be free from undue influence, duress or restraint.

The first trial was before Judge Van Brunt, but he would not give the case to the jury, but directed a verdict for defendant. An insane delusion affecting the provisions of a will must invalidate it. Now, can a belief in Spiritualism and communications from the so-called spirit land be considered an insane delusion ? A very safe rule in medico-legal trials of this sort is the following : *When a person entertains a belief opposed to the general experience of mankind, and incapable of being verified by human means, and such belief leads the person to disregard the ordinary obligations of duty and affection, it is to be hoped that a will based upon such conditions may never stand.* Any alleged religious belief that leads a testator to commit wrong and injustice may safely be set down as a delusion.

CHAPTER VIII.

MENTAL RESPONSIBILITY IN CRIMINAL CASES.

AT the present day medico-legal cases are becoming very frequent in which it is necessary to ascertain as to the insanity of a person accused of a criminal act in its relation to his civil capacity and responsibility for criminal actions, and also as to feigned or concealed insanity. It becomes, therefore, a very interesting question what test of insanity the law should recognize as a valid defense in criminal cases. This question—although one which it seems difficult to settle satisfactorily, and which judges, lawyers, and medical experts are constantly disputing about—assumes, every day, greater interest and wider significance, owing to the increase of insanity in our country, disproportionate to the increase of population, which has taken place during the past twenty years, and which will continue to take place. Without inserting dry statistics, it is sufficient to say that a comparison of the increase of population, from 1850 to 1870, with the increase of the number of the insane during the same period, reveals an increase of insanity over that of population of about twelve per cent.

The increase of insanity among our own population is due largely to a change from a vigorous, well-balanced organization to an undue predominance of the nervous temperament which is gradually taking place in successive generations. The educational pressure on the young, to the neglect of physical exercise, the increasing artificial and unnatural habits of living, the great excitement and competition in business, are all tending to induce and multiply nervous diseases, many of which must terminate in insanity. These causes and the evils resulting from them are propagated by the laws of inheritance in an aggravated and intensified form. Insanity is also appearing gradually at an earlier age than formerly.

This is due largely to the great mental activity and strain upon the nervous system that appertain to the present age and state of civilization, and which tend to a rapid decay of the nervous system. With many persons it is but a step from extreme nervous susceptibility to downright hysteria, and from that to overt insanity. The question of mental responsibility in its relation to criminal cases is one of great interest, and presents a wide field for study and investigation. The facts of criminal psychology have led the writer to regard the impulse of criminal natures in the light of natural laws, and there is, beyond all doubt, an anthropological change which lies at the foundation of criminal propensities.* There is a deficient cerebral organization which lies at the foundation of these criminal natures, which occasions the disposition to an abnormal moral constitution. The dislike of work and the love of enjoyment are impulses which, when combined, lead especially to crime, when that ethic constitution or development is wanting which is necessary to the foundation of a powerful feeling of what is right. A further fundamental element, which stands in psycho-physical contrast to dislike of work, is an excessive physical consciousness of strength, which leads to arrogance and thereby to the pleasure of misusing strength against the weak. This impulse leads to the love of bullying, cruelty, and murder, if the higher intellect is absent which should turn the feeling of strength in a right direction, and there is also absent a complete ethical consciousness which should prevent

---

* Benedikt has shown this conclusively.

misuse of power. This ethical weakness may be congenital, as has been remarked, or it may arise from
deficient education.

In the domain of vices we meet with a peculiar condition of the central nervous system, which results in a
temporary criminal impulse, returning with a certain
regularity. Such criminals are temporarily seized with
the deepest remorse and are fortified with the best reso-
lutions. They behave for a time in the most exemplary
manner, until they relapse again, which relapse is unanimously attributed by them to an irresistible impulse.
This state of *moral epilepsy* is of great significance in the
psychology of crime, as a physiologist is led to institute
a comparison between such cases and several states of
disease in which a peculiar type is observable, consisting
in the fact that attacks of illness of more or less duration alternate with more or less long and generally for a
time preponderant healthy intermissions. In a broad
sense, one may designate all these pathological states as
epileptiform, hence the term "moral epilepsy," which
has been adopted above. Leaving this interesting question of the psychology of crime, we would ask if the
true basis for jurists to proceed upon is not *the
protection* of the existence of normal persons against the
ethically degenerate? And the necessary degree of this
protection is, most certainly, an essential measure for
the severity of the punishment. The first trial of note
where there was the question of insanity advanced was
in 1723, when the trial of Arnold for shooting at Lord
Onslow occurred. Although it was shown that Arnold
had been of weak understanding from his birth and that
he was doubtless insane, the jury brought in a verdict of
guilty, and Arnold would have been executed had it not
been for the intercession of Lord Onslow.

The language of the charge to the jury in this case
was in conformity to the rule laid down by Lord Hale,
that partial insanity does not excuse a person from the
consequence of his act, and that a total deprivation of

reason can furnish such an excuse. In the year 1800 the celebrated trial of Hatfield for shooting at the King, in Drury Lane theatre, excited much interest. Although it was proved that in 1793 Hatfield, who was a dragoon, had received a number of severe wounds which had caused partial insanity, so that he was dismissed from the service, and since that time he had had periodic attacks of insanity, and had been confined as a lunatic, the prosecuting attorney laid down the established rule that a total absence of memory and understanding could alone shield the prisoner from punishment, and appealed to the jury for a conviction on that ground. It was only through the brilliancy of the advocate (afterward Lord Erskine) that the prisoner was acquitted. This trial had a good effect upon the judiciary, as in the year 1812, in the trial of Bellingham for the murder of Spencer Percival, Lord Mansfield laid down the law that the capability of distinguishing between right and wrong was the test for determining the prisoner's responsibility, thus discarding the old theory of an entire absence of all mental power and substituting this in its place. Afterward the theory of a general knowledge between right and wrong was modified, and the element introduced that the prisoner must know the difference between right and wrong at the time of and with regard to the particular act for which he is on trial in order to render him responsible, and this test has been preserved to the present time. In the early history of our own country the same barbarism in the treatment of the insane prevailed which darkens the pages of English history. In Governor Winthrop's "History of New England" the case of Dorothy Dalbye is mentioned. She was executed for killing her child. She was, beyond all doubt, an insane woman, but this fact was not recognized by Governor Winthrop, who says of her: "She was so possessed with Satan that he persuaded her by his delusions, which she listened to as revelations from God, to break the neck of her own child, that she might free it from future misery." Such was the ignorance and prejudice of the early history of our country.

We are at the present day very far from a correct understanding of the workings of the insane mind, for in the recent trial of Scannall, the law was laid down as enunciated by the Court of Appeals in 1865, in the case of Willis v. The People, which held that a person was not insane who knew right from wrong, and that the act he was committing was a violation and wrong in itself. This theory of right and wrong is utterly inadequate to meet a large class of cases. There are certain cases familiar to all specialists in insanity, which suffer from impulsive insanity with a homicidal or suicidal monomania. These patients, without appreciable disorder of the intellect, are impelled by a terrible *vis a tergo*, a morbid, uncontrollable impulse, to desperate acts of suicide or homicide. These patients are often fully aware of their morbid state, appreciate perfectly the nature of the act toward which they are impelled, and feel deeply the horror of their situation, and yet if not prevented by restraint will inevitably commit acts of suicide or homicide.

A very remarkable case was under the care of the writer, of a man who would at stated times acknowledge that he felt an irresistible impulse to kill some one, and would voluntarily enter an asylum and remain there until this morbid impulse had passed away, which was generally a period of one or two months. He has often told the writer that his life was made miserable by the idea that at some time this overwhelming impulse would come upon him so suddenly that he should commit some desperate homicidal act, but is not prepared to voluntarily incarcerate himself in an asylum for life, as his lucid intervals sometimes lasted for months at a time. The law, as laid down at present, would not decide this man to be insane, as he fully appreciates the difference between right and wrong, and the nature and consequences of any homicidal act that he may in the future commit. Such cases, which are not at all uncommon, serve to show what fearful injustice may be done under the name of justice, when the conclusion is based upon a metaphysical test which is proved by medical observation to be false in its application to the unsound mind. There is still another form of insanity denominated "moral insanity," in which the intellectual faculties are intact, no delusions or hallucinations existing, but where the moral sense seems utterly obliterated. Such persons have no true moral feeling. This is disorder of the mind produced by disease of the brain, and it is an unquestionable form of insanity, as it often precedes other forms of insanity, in which intellectual derangement is well marked, as acute mania or general paralysis. In some of these cases there is a modified responsibility, the degree of such responsibility being determined by the particular circumstances of each individual case. One difficult but important question to be solved is the civil and criminal responsibility of women who plead insanity before courts of justice, and who are often afflicted with kleptomania, pyromania, or who are infanticides, as a result of sexual trouble and disease of the pelvic organs. Such women, under all reasonable conditions, are entitled to the benefit of the doubt, because of their defective mental integrity, caused perhaps by pregnancy or by the subsequent emotional excitement attending parturition, which intensifies the cerebral disorder in a brain already morbidly active.

With women, extreme nervous susceptibility readily lapses into insanity. "In the sexual evolution, in the parturient period, in lactation, strange thoughts, extraordinary feelings, unreasonable appetites, criminal and suicidal impulses may haunt a mind at other times innocent and pure. It is probable also that young unmarried women, guilty of killing their own new-born offspring, are so distracted by conflicting feelings, sharpened to morbid acuteness by the great physiological movement of parturition, as to be hardly responsible for their acts." We come now to the question of the *diagnosis of insanity*.

In making an examination of a person accused of crime, and in whom insanity is suspected, the person should be visited by the medical examiner, who should draw him into a pleasant conversation, and inquire as to previous attacks of insanity, hereditary history, then into any predisposing causes of insanity, such as intemperance, vocation, habits, etc., which may have operated in the production of insanity. Also as to injuries of the head or spine which may have occurred, sunstroke, etc. The nervous system should then be examined for the existence of any such diseases as paralysis, epilepsy, catalepsy, or hysteria. The different senses, beginning with sight, should be examined, and in this way it may be discovered if there are hallucinations or illusions pertaining to any of the senses. A great many cases are on the border line which separates sanity from insanity, and it often requires the nicest discrimination to determine whether such a patient has passed this border line. The writer would suggest a series of eight questions, which, if adopted by jurists in criminal cases, would prove a most efficient and just test as to the existence of insanity in any given case, viz. :

1. Have the prisoner's volitions, impulses or acts been determined of influenced at all by insanity, and are his mental functions, thought, feeling and action, so deranged, either together or separately, as to incapacitate him for the relations of life ?

2. Does the prisoner come of a stock whose nervous constitution has been vitiated by some defect or ailment, calculated to impair its efficiency or derange its operations ?

3. Has the prisoner been noticed to display mental infirmities or peculiarities, which were due either to hereditary transmission or present mental derangement?

4. Has the prisoner the ability to control mental action,

or has he not sufficient mental power to control the sudden impulses of his disordered mind, and does he act under the blind influence of evil impulses, which he can neither regulate nor control ?

5. Has the act been influenced *al all* by hereditary taint which has become intensified, so that the morbid element has become quickened into overpowering activity, and so that the moral senses have been overborne by the superior force derived from disease ?

6. Was the act effected by or the product of insane delusion ?

7. Was the act performed without adequate incentive or motive ?

8. Does the prisoner manifest excitement or depression, moody, difficult temper, extraordinary proneness to jealousy and suspicion, a habit of unseasonably disregarding ordinary ways, customs and observances, and habitual extravagance of thought and feeling, and inability to appreciate nice moral distinctions ; and, finally, does he give way to gusts of passion and reckless indulgence of appetite ?

Some or all of these are found generally in connection with transmitted mental infirmity. It may be argued that these mental defects signify not mental unsoundness, but human imperfection. Certainly if we take these manifestations, any one of them singly and alone, we cannot claim such a one as invariably an indication of insanity ; but, on the other hand, under certain circumstances, each one of them may be an unmistakable sign of insanity, or rather of a morbid cerebral state which may readily lapse into insanity. The disappointments and calamities of life obviously act with greater effect upon an unstable mental organization, these causes of disturbances meeting with a powerful co-operating cause in the constitutional predisposition. Sometimes a crime even when there have been no previous symptoms to indicate disease, marks the period when an insane tendency has passed into actual insanity, when a weak

13

organ has given way under the strain put upon it. There is a class of persons with a peculiar nervous temperament who inhabit the borderland between crime, and insanity, one portion of which exhibit some insanity, but more of vice, and the other portion of which exhibit some vice, but a preponderance of insanity; and it is very difficult to form a just estimate of the moral responsibility of such persons, especially when we reflect upon the fact that moral feeling is a function of oganization, and is as essentially dependent upon the integrity of that part of the nervous system which ministers to its manifestations, as in any other display of mental function.

The writer has met with cases in which, as a result of parental insanity, there has been a seemingly complete absence of moral sense and feeling in the offspring, and this has been a true congenital deprivation, or a moral imbecility so to speak; of course such children can hardly fail to become criminals. In this connection, it is interesting to note that moral degeneracy often follows as a sequence upon disease or injury to the brain. A severe attack of insanity sometimes produces the same effect, the intellectual faculties remaining as acute as ever, while the moral sense becomes obliterated.

When such persons are acquitted, on trial, of a criminal act on the ground of insanity, they should be remanded to medical custody, and should never be set at liberty until the medical superintendent of the asylum deems them fully recovered; but the commonest justice plainly indicates that such custodial restraint be of a medical and not of a penal nature. It is a very difficult thing for the laity to recognize how sane a person may be who, all the while, has a greater derangement than was ever suspected until something happens to elicit the evidence of it, such as an attack of illness or severe mental strain, and some unconquerable impulse seizes him, and some homicidal or suicidal act results, to the great surprise of everyone.

In the same manner inebriety often appears in **maturity**

as a result of ill-health, mental shock, etc., and it becomes an interesting question as to the degree of moral and criminal responsibility which attaches to inebriates, as inebriety often depends upon an abnormal organic development of the nervous system that has descended from generation to generation, gaining in intensity until it manifests itself in active inebriety, and there must certainly be a modified responsibility when homicidal or suicidal acts are committed during periods of such abnormal cerebration. In such cases a criminal act may be committed in consequence of cerebro-mental disease, without any apparent lesion of the perceptive and reasoning powers. In these cases, also, the mental disorder is of a sudden and transitory character, not preceded by any symptoms calculated to excite suspicion of insanity. It is a transitory mania or sudden paroxysm, without antecedent manifestation, the duration of the morbid state being short and the cessation sudden. In these cases the criminal acts are generally monstrous, unpremeditated, motiveless, and entirely out of keeping with the previous character and habit of thought of the individual. Such attacks are transient in proportion to their violence, and transition occurs on the completion of the act of violence. There is an instantaneous abeyance of judgment and reason, during which period the person is actuated by mad and ungovernable impulses.

Closely allied to this state of which I have been speaking, is that peculiar psychological state — the trance state — which also occurs in inebriety. There has been very little medical study of these cases, although they are of great medico-legal importance. Crimes committed in this state are purposeless, and there exits *no* recollection of them in the mind of the sufferer. By "trance state," I mean a state where there exists loss of memory and consciousness for a time, varying from minutes to days, the patient giving no evidence by his acts of his real condition, and very likely attending to all of the duties of his business in a quiet, mechanical way.

The mind may, however, in this trance condition, act in *unaccustomed* lines of thought and action, and, in certain cases, the *criminal* impulse may dominate the mind. As a rule, in these trances, it is probable that the mind acts, as before, with the same discretion, although the person himself can give no account of what has happened during this mental blank, during which the mind acts automatically. During this state a person may get into a dangerous mental condition, in which impulses of every description may take possession of and control his actions. It is a condition of irresponsibility. In these cases there is generally a neurotic constitution inherited from the ancestors, and a careful examination, which should never be neglected, will generally reveal either intemperance, insanity, or phthisis in the ancestors. The great diagnostic point which I would insist upon is that there is absolutely *no recollection* of what happens in this trance state, and this want of memory cannot be successfully feigned so as to deceive a careful expert in inebriety and insanity. When a person, either a secret or an open inebriate, commits an unusual or criminal act (not during intoxication), and retains no recollection of the event, he should be most carefully examined for the existence of the trance state, which, if it can be proven, markedly lessens the responsibility of his crime.

The impulse to crime in these cases may develop in different ways. There may be suicidal or homicidal impulses, or buildings may be burned, or sexual assaults may be made, or acts may be apparently purely malicious. In all these cases medical care is plainly indicated. Persons unacquainted with this state may reason that because these crimes are committed in a way and manner perfectly cool and free from excitement, that they are evidences of a sane mind ; but they are greatly mistaken, as it is the rule, and not the exception, to find these deeds performed in a cool, quiet manner, with no excitement. The deeds, however, are, to the person affected, *an unusual course of action*, utterly at variance with the previous character and habits.   It is not right for inebriates who commit sudden, purposeless crimes to receive the full measure of punishment at the hands of court or jury, without a proper study of their case by one who has studied these cases intelligently.   Proper medical study would enable courts and juries to understand the mental conditions which cause these motiveless and purposeless crimes in inebriates.   His act is not a vice, and you can neither assume his perfect sanity nor a capacity to reason clearly. The person in this trance state is not cognizant of his acts, neither, as I have said, does he retain *any* recollection of them afterward.   There is absolute irresponsibility, and it should be made clear to both court and jury.   The idea in these cases should not be to try to work on the sympathies of the jury, but to show them the existence of this trance state and the irresponsibility arising from it.   In these cases we must prove that inebriety, as a disease, exists, and that it has affected intellect, manner, temper, disposition, habits and character, and then that the trance state has supervened.

We will consider, finally, the medico-legal importance of epileptiform attacks, which may be partial in character, and which may not reach convulsive activity except so far as the mind is concerned.   These attacks always display periodicity, and after the paroxysm there is an intermediate stage, during which, in most cases, the person remains in a confused state, perhaps for some hours, and is apt subsequently to retain only a vague and general notion of the preceding events.   Thus, in a homicide by shooting, the murderer would be likely to be roused by the sound of the pistol-shot and to remember it, although he would not very likely remember the altercation at all, or what passed between himself and his victim.   A case occurred recently, of considerable interest from a medico-legal point of view, in which a murder was committed during an epileptiform seizure, which was the result of a previous sunstroke, the immediate exciting cause being an attack of illness and the taking of a small quantity of alcoholic stimulus, which, it is well known, acts as a poison upon persons who have been sunstruck. This state of what, perhaps, I may not improperly call moral epilepsy, in which the man was of whom I shall presently speak, is a morbid affection of the mind centres which destroys the healthy co-ordination of ideas and occasions a spasmodic or convulsive mental action. The will cannot always restrain, however much it may strive to do so, a morbid idea which has reached a convulsive activity, although there may be all the while a clear consciousness of its morbid nature. The case just referred to had complained of pains in the head and sleeplessness, which had displayed marked periodicity, and which had been accompanied with great irritability of temper, excited by trifles and seemingly unconnected with personal antipathies. As has been previously stated, the person alluded to had been suffering from quite a severe illness, and, after taking a small quantity of alcoholic stimulus, went out to walk. He met a friend with whom he had been familiar for years, and a discussion arose as to the respective merits of certain politicians, when, the discussion becoming excited, the man pulled out a revolver and shot his friend. He then went, in a confused and dazed state, and sat for some hours on a dock near a river, and subsequently went home and burst into tears, and informed his wife of the sad occurrence and gave himself up at the police station. There was no simulation of insanity by pretending to be incoherent or by strange actions, and no attempt, either on the part of himself or wife, to pretend that the act was an insane one. There was, however, a total blank in the prisoner's mind respecting the events preceding the pistol-shot, which seemed to have aroused his attention at the time, and he had no recollection of the fact that he had sat on the dock for some time afterward, as he was seen to do.

I was consulted by Judge ——, who appeared for the defense, and, upon ascertaining the prisoner's previous history, gave it as my opinion that there had existed, for

months previous to the occurrence, a profound moral or
affective derangement, which, from its marked periodicity, was evidently epileptiform in character, and that
the sudden homicidal outburst supplied the interpretation of the previously obscure attacks of recurrent
derangement.   There had evidently been induced by the
sunstroke in this case an epileptiform neurosis, which
had been manifesting itself for months, chiefly by irritability, suspicion, moroseness, and perversion of character,
with periodic exacerbations of excitement, all foreign to
the man previous to the attack of sunstroke.   It is well
known among specialists in insanity that this epileptiform neurosis often exists for a long time in an undeveloped or masked form, and that this neurosis is, moreover,
connected with both homicidal and suicidal mania.   Such
attacks are often noticed to occur periodically for some
time before the access of genuine epilepsy.   I have often
witnessed, in cases under my care, abortive or incomplete
epileptiform attacks, where there were no convulsions and
where there was no complete loss of consciousness.   I
have noticed in such cases, either a momentary terror,
slight incoherence, a gust of passion, or a mental blank,
the patient perhaps stopping in the middle of a sentence.
The patient would then be himself again, quite unconscious of what had happened to him.   Accompanying
this confusion of ideas may be, as I have remarked,
instantaneous impulses, either of a suicidal or homicidal
nature, and acts of a destructive nature.

Owing to the writings of Hughlings Jackson, Maudsley, Russell Reynolds, Hammond, Trousseau, Falret,
Esquirol and others, epileptic vertigo is a recognized disease.   There is abundant testimony to show that during
such seizures persons may perform actions, and even
speak and answer questions automatically.   There are
numerous examples in the works of the above authors,
proving that in an unconscious condition persons can progress from odd or eccentric actions to deeds of violence,
suicide or murder—being unable to remember the circum-

stances afterwards, and, therefore, irresponsible for their actions. This class of patients I have always found irritable, easily excited, very emotional without adequate external cause, easily losing their train of thought, and often unable to collect or fix their thoughts. Such cases have told me that they felt impelled to strange and violent acts by some power which they could neither understand nor resist. Such patients may entertain delusions of fear and persecution, and commit criminal deeds as a result of such delusions. When such cases, in their terror or distress of mind, commit some violent deed, they either experience immediate relief, as was the case with one patient under my care, who was only relieved by breaking out a pane of glass, when his paroxysm would subside, or they continued in a state of excitement, unconscious, or very imperfectly conscious of the gravity of their acts. When they become conscious again their memory is apt to be very uncertain as to preceding events.* Griesinger says: "Individuals hitherto per-

---

* The reader must distinguish between such cases as the above and cases of true *homicidal mania*, which latter, may be classified under two heads: 1. Those cases in which there is no marked disorder of the intellect, examples of emotional insanity. 2. Those in which such disorder is more or less apparent. The former class may be subdivided according as there is or is not evidence of premeditation and design. In the latter class we include cases marked by deficiency of intellect, as idiocy, imbecility, and a degree of mental feebleness not amounting to either of these states, while other cases in the second class are rather indicated by a state of exaltation, shown by delusions or hallucinations which may constitute the motive. When any person says that they had not the slightest motive for a homicidal act, *I always suspect masked epilepsy,* even if there is no proof of it, although there is, of course, a morbid, overpowering impulse to take life, without intellectual disorder and with intact perception and reasoning powers. There has been *petit mal* in nearly every case of transitory mania that we have knowledge of. Griesinger has truly said that we may have a morbid mental state which may present no external manifestation. Legrand du Saulle's case of a young man, "Theodore," of twenty-six years, who assassinated two men without premeditation, without motive, and without apparent excuse, was undoubtedly a case of epilepsy complicated with transitory mania. Castleman, Devergie, Calmeil, Tardieu, Marcé, have

fectly sane and in the full possession of their intellects are suddenly and without any assignable cause seized with the most anxious and painful emotions, and with a homicidal impulse as inexplicable to themselves as to others.†" Maudsley says: "Let it be borne in mind, then, that there are latent tendencies to insanity which may not discover the least overt evidence of their existence, except under the strain of a great calamity, or of some bodily disorder, and that the outbreak of actual disease may then be the first positive symptom of

---

all reported cases of this nature. A young man of education placed himself voluntarily under our professional care, acknowledging impulses to commit homicide, and said that he did not like to have razors or knives around him for fear that he should be unable to resist these impulses. "I never feel sure of myself," he told me. He evidently felt himself to be irresponsible, and expressed himself forcibly to me on this point. I urged him to voluntarily place himself in a well-regulated asylum for the insane, but neither himself or his family would acquiesce in this measure of preventive medicine, and he is still at large. He said he had received a blow on the head and had also suffered from heat prostration to the extent of insensibility, and complained of cephalalgia. I have had an opportunity of studying another case of what I think is pure moral insanity, where there is an utter moral insensibility, and where the whole channel of thought runs in the direction of how to successfully commit homicide. There are sexual perversions in this case. There is marked sleeplessness, but the perceptive and reasoning powers are very good. There is premeditation and cunning design, and a knowledge of right and wrong. This patient has also suicidal impulses.

† On *mania transitoria*, see Kraft-Ebings' "Die Lehne von der Mania Transitoria fur Aerzte und Juristen Dargestelt," Erlangen, 1865. Article by Dr. S. T. Clarke in the "Am. Journal of Insanity," January, 1872, and one by Dr. Jarius, July, 1869, on " Mania Transitoria." Also Castelman and Devergie, 1851 and 1859, respectively advocate "instantaneous insanity." They think there is generally heredity. Calmeil and Tardieu also believes in this form of insanity. A homicidal act *alone* may constitute sufficient evidence of the insanity of the homicidal. *There are, beyond all doubt,* transitory aberrations of the intellect, which produce destructive and homicidal impulses, with no prior history of mental disorder, unless, perhaps, of some degree of mental irregularity.

unsoundness." The question as to the degree of mental responsibility attaching to such cases is one of great interest to psychologists and also to jurists, and one to which, it is hoped, in the future, much more attention may be directed than has been given to it in the past.

CHAPTER IX.

THE LEGAL RELATIONS OF STATES OF UNCONSCIOUSNESS, SOMNAMBULISM, CATALEPSY, ETC.

THERE is great interest and importance attaching to this subject, but we are accustomed to very vague and undefined ideas respecting it. My effort in this chapter, aside from treating of the diseases of somnambulism and catalepsy, will be to contribute, in however slight a degree, to the knowledge of the more exact relations of the human mind and of human acts to responsibility.

We have been accustomed to regard the partial interference with sensibility and mobility, and the resulting limitation of will in trance, trance-coma, somnambulism, catalepsy, and epilepsy, as curious physiological states rather than as diseased states of the nervous system requiring medical treatment, and also seriously affecting mental and legal responsibility. It is only in the most perfectly balanced minds, where there is an accurate balance between the subjective and the objective faculties, that consciousness is never impaired, and where there are no breaks in the continuity of perception and memory, during which time the connection of the individual with the thing done or said is no longer reliable or distinct. Even in men of the strongest mental calibre such obliviousness sometimes occurs. In whatever these intercurrent spaces of a non-existence may have originated, whether from unfettered determination, or the idle wandering imagination, or from the brain-wasting following moral or intellectual hard work, it is certain that occasionally they pass beyond the power and in defiance of the will, and should be classed under the head of morbid nervous affections, if not with actual mental disorder.* Men of lofty intellect and vigorous and acute minds, by excessive and continuous application, overtaxing their attention and introspection, and confining the exercise of their intellect and memory within a narrow range, weaken their observant powers, and by concentrating their minds upon particular objects, produce oftentimes grave disorders of the nervous system. Sir Joshua Reynolds and Sir Isaac Newton are prominent examples of attention so long fixed and contemplation so intense as to render them entirely oblivious to self and surroundings and to disturbances in perception. After Sir Joshua Reynolds had been for hours occupied in painting and walked out into the street, the lamp-posts seemed to him to be trees, and the men and women moving shrubs. He had fixed his attention for such a length of time on the picture before him that he could not direct it to other objects of sensation.

A very remarkable instance of forgetfulness and absence of mind occurs in the biogrophy of Hookham Frere, the scholar and man of letters, who, handing the Countess of Errol to supper, drank the negus he had prepared for her, and altogether forgot the object of their visit to the dining-room ; and who, on the day of his marriage with the same lady, had no recollection, until the evening, that he had promised to accompany his bride to the country, having occupied the intervening time in reading his poem to his publishers. It is a psychological fact, that after the attention has been for a great while intensely fixed upon particular objects, the person cannot direct it at will to other objects of sensation. Any occupation or exercise which narrows the scope of intellectual exercise, which nullifies the influ-

---

* This was clearly laid down in the London Journal of Psychological Medicine by Dr. Forbes Winslow some years ago.

ence of the emotions or contracts the mental forces, is directly prejudicial to mental health, by giving undue prominence to certain faculties and allowing others to fall into disuse and apathy, overstimulating some regions of the brain and probably producing undue or defective nutrition in certain parts of the brain. It is the peculiarity of all these states, when they are not merely temporary effects of overwork, that they essentially consist in such consequences of bodily or mental degeneration as, robbing the thinking part of our nature of its nobler endowments, leave it in the impoverishment of an appetite, a peculiarity, or a single all-embracing thought. The morbid element consists chiefly in fixedness — in the inability of the will to substitute another train of reflection or perception. The will, though feeble and fickle, is not entirely extinct, as it serves to guide in the direction of the predominating, if not constantly permanent, notion or incentive. It seems to me that we must class such states in the same category as the preoccupation of the insane, whose disordered imaginations can admit of nothing but the present ruling impulse, and with the absorbing and exclusive anguish of the melancholiac. In all these cases there exists, in different degrees, a suspension of consciousness. These spaces of non-existence are on the border line which divides sanity from insanity. The morbid states of the brain which may be induced are exemplified in a marked degree in the lives of ascetics and ecstatics, and in those whose intense devotional feelings, as in the convulsionaries and Brahmins, extend, for the time, to enfeeblement of volition and to diseased functions of motivity and sensibility, manifested in violent convulsions and complete loss of sensation.

In *Somnambulism*, the first of the morbid states which I propose to consider, we have, as the constant and unvarying state, a morbidly profound sleep, in which "the sceptre of reason is surrendered to a physically directed fancy." It is due, probably, either to an overloaded stomach pressing on the solar plexus of nerves, producing a partial paralysis in the coats of the arteries, and so in the circulation of the brain ; sleeping with the head too low, and strong mental emotion.   It is a peculiarity of somnambulism that even after the removal of the cause, the habit, once established, is apt to remain. It is most frequent in youth, and about the age of puberty.   In the states of unconsciousness accompanying somnambulism, the senses are awake and preternaturally alive.   The muscles are regulated, and regulated, too, with wonderful precision and power.   There is a purpose, and there is a co-ordination of acts for its accomplishment ; but consciousness is still asleep, and memory retains no record of the transaction, although it may have been prejudicial in the highest degree to the interest of the actor or of others.   In many states of unconsciousness the mind is forced to think or feel in a particular way, and is forced to instigate certain deeds in flagrant opposition to its ordinary character and tendencies, and in utter disregard of the promptings, or of the resistance of other motives and considerations. There is a very close relation between acts committed during states of unconsciousness and mania transitoria, epileptic paroxysms, and the irresistible impulses of insanity.   They have, in common, irresistibility, suddenness and rapidity.   They are alike unannounced and of short duration.   They are alike characterized by the exercise of free-will being fettered or perverted, and there are, undoubtedly, distinct morbid conditions in all of these different states.   If we examined with sufficient care, cases in which unconsciousness occur, I feel quite sure we should discover the prodromic signs which have been observed to usher in other species of the neuroses. Somnambulism may be hereditary, but it is not inconsistent with fair health.   It is apt to become periodical, patients having attacks once a week, fortnight or month.

The *treatment of somnambulism* consists in preventing the very deep sleep, in which the phenomena of somnambulism are exhibited.   The patient should be awakened once or twice a night before the phenomena begin to appear; soon after retiring, and again after four or five hours' sleep, will usually answer.   Patients should dine in the middle of the day, and while taking care that all meals should be light and digestible, we should be particularly careful not to overload the stomach at night.   The use of electricity and nerve tonics to bring up the general health to the highest point are indicated. Friends should be cautioned not to awaken the patient while walking, as the fright may act prejudicially.   He should be quietly put back to bed.   The head should be well propped up by pillows, and too great a weight of clothes must be avoided.

Catalepsy.   I find an excellent definition of catalepsy in Dr. Boerhaave's aphorism, published in 1755.   He graphically describes it as "that disease in which the patient becomes of a sudden unmoved, void of feeling, and retains the same posture and action of all the parts of his body which it was in when the disease seized him first."   It is a disease of central innervation of the nervous system, and may be accompanied by or accompany many forms of insanity.   In a cataleptic paroxysm, the state of unconsciousness is characterized by the limbs of a patient remaining in the position in which the patient had placed them before the inception of the paroxysm, or in which any by-stander may place them during the paroxysm.   Consciousness and sensibility are entirely suspended.   Catalepsy may accompany insanity and chorea, and many of the neuroses.   If death is simulated, the existence of mucular contractility under the Faradic current, and also the dark eschar of the cautery, are tests which may be applied to determine life.   The patient's will is powerless to act during the paroxysm, by reason of the muscular contraction induced by excitement of the motor nerves, proceeding from the spinal cord.   The paroxysm is preceded by dizziness, headache and a very irritable state of the general nervous system, and begins very suddenly.

There is apt to be a vague uneasiness and sleeplessness. A patient of mine presented the following typical symptoms and manifestations of a cataleptic attack in my presence. The lady in question, who was from North Carolina, while in the act of conveying a morsel of food to her mouth, became suddenly rigid and pale, the arm being arrested in its passage and being immovably fixed, with the fork in the hand a few inches from the mouth. The whole body was as motionless as if the patient were carved out of stone. The eyes presented a widely opened, staring condition, and consciousness and sensibility were entirely suspended. Respiration could not be detected and the pulse-wave could not be felt at all. In about four minutes the patient sighed deeply, made a full inspiration, and resumed her meal, quite unaware of what had happened to her. The cataleptic trance may last for some hours possibly, and in extreme cases may last for days. Patients remember nothing of an attack or what transpires during the trance-like state.

Catalepsy, although not necessarily connected with insanity, is, I think, very often dependent upon an insane temperament or neurosis. It has been stated that catalepsy is generally a complication of hysteria, but the results of one hundred and forty-eight cases collected by Dr. Puel, in which sixty-eight occurred in males, would seem to disprove the assertion. An interesting case of this rare disease was reported by Dr. S. S. Cornell, of Toledo, Ont., not long ago. The catalepsy came on after the second confinement, before which the patient was very nervous. After the confinement there was a chill, followed by sharp febrile action, with pain and tenderness over the region of the uterus. There was some delirium and suppression of lochia. This condition, however, disappeared, but was followed by a cataleptic state, which I give in the doctor's own language : " Now comes the sequel. The patient passed the next forty-eight hours most beautifully, except on the night of the 30th she could not sleep ; otherwise the nurse thought she was doing extremely well. A peculiar change was soon discovered taking place with the patient ; her acuteness of hearing was extremely great : could hear and reiterate the sentiment of persons in the adjoining room, who conversed, as they declared to me, in a low whisper, and that they conceived it impossible for a person to hear a word whispered six feet from them ; yet this patient, at a distance of twenty feet or more, with closed doors, could tell the sentiments exchanged. This was done several times, and finally the patient called her husband to her, kissed him ; then called her little boy three years old, and her infant, kissed them and then bade her friends adieu. This procedure of my patient awoke a deep interest in the minds of the nurse and friends, who now became alarmed. The nurse persuaded the friends to leave the room to her and the patient, as she thought after a little Mrs. H. would fall into a repose ; but instead of sleep, our patient lay speechless and motionless, with eyes staring wide open, no signs of respiration ; they opened her mouth to see if she would swallow, but in vain ; her lower jaw remained depressed as the nurse had left it. Attempts were now made to rouse her by calling loudly in her ear, but to which she paid no attention. They thought her dead, and that it was useless to send for medical aid ; thus passed away twelve hours, when her husband dispatched a messenger for me. When I arrived and entered the room, I was shocked to see what struck my fancy to be a waxen figure or a frozen corpse in lieu of my former patient. There she lay with under jaw depressed, eyes staring and wide open, without winking, the pupils a little dilated, skin cool, almost the feel of a corpse before stiffening, pulse 122, feeble, no sign of respiration. In examining the pulse I raised the arm to see if that would cause any difference in the pulse. There it remained for nearly an hour, when I put it down by her

14

side. There was but little resistance offered to any change of her limbs or person ; but whatever attitude a limb was placed in, there it remained. I now brought her under jaw up to its place and it remained. I was importuned to do something for the patient. What to do was with me a paramount question. The thought occurred to me that I might administer an enema of strong solution of assafœtida, which I did to the amount of a quart ; and this was very easily done as there was not the slightest resistance. Still the patient lay as lifeless as ever for about an hour, when a few slight convulsive movements were observed, and she roused to consciousness. She looked about her, asked what had been done with her corpse, as it appeared to her that her friends desired her to remain for a season, but her judgment dictated to her to again depart and take her infant with her. I gave her several doses of assafœtida, fluid extract of valerian, beef tea, etc. She now desired to be left alone, as she said she had an important duty to perform, and the presence of persons, however nearly related, was detrimental to her welfare. She was satisfied for me to remain with her alone, as she said, "from the days of antiquity, deference had always been paid first to the priest and then to the doctor." She remained quiet for, in all, a period of six hours, taking beef tea, valerianate of ammonia, assafœtida and bromide of potassium. Soon she drew the sheet over her face, and then placed her arms over her chest and lay straight in bed ; she lay so quiet and still that I felt induced to remove the sheet, when, as I had feared, I found her in a second trance (?). Eyes wide open, pupils a little dilated, but would contract under the influence of strong light ; skin cold, of a death-like feel, no rigidity of the muscles ; pulse 112 and very feeble ; not the first sign of respiration, no movements of the nostrils. I now lifted her body up to an obtuse angle with her lower limbs, I next raised one arm and then the other, and in this position I left her for several minutes. I now stepped back, gazed upon my patient, who, in a semi-sitting posture, with staring eyes, with outstretched arms and a lifeless appearance, appeared as though a corpse had thus been placed and left to stiffen.

I then laid her down upon the pillow, raised her body up, having her head on the pillow in the attitude of opisthotonos, and thus she remained ; after a period of twenty minutes, I gave her a slight push and she fell on her left side with her body still having the same curve. I now straightened her out in bed, spoke loudly to her several times, but no response. I again repeated the assafœtida injection, containing ol. terebinthinæ. To please her friends, I tried several times to have her swallow, but all to no purpose. I held to her nose strong aqua ammonia, which affected her in no perceptible way. In this state she lay about eight hours ; when consciousness returned she related what she saw while in the other world. This time she was not so composed and tranquil as when she came out of the first trance (?) Her symptoms now assumed more the character of hysteria, her limbs were affected with convulsive twitchings, and she screamed loudly without giving utterance to any cause for so doing. When she went into the second state of mental abeyance, my views were, as soon as consciousness returned, that she should be brought under some powerful anæsthetic, whereby her mental state might recuperate. Whether this should be produced by chloroform, ether or hydrate of chloral was not fully settled in my mind. I therefore sent for Dr. Addison of Farmersville, who arrived just after her imperfect return to consciousness. It was decided at once to give her hydrate of chloral, of which she took seventy grains in the space of an hour, after which she fell into a profound sleep and did not awaken for twelve hours. Her convalescence then commenced."

These cases are of interest to the practitioner, although comparatively rare, as the cataleptic paroxysms or fits annoy or disturb the patient's mind, lest they should come on while travelling or away from home and friends. Although the fits generally last but a few minutes, they may possibly last for several hours or even days. The chief indications for treatment are to improve the general nervous tone by nerve tonics and electricity; induce the patient to lead an outdoor life, eat regularly, avoid rich indigestible food; to retire early; and, if the patient complains of a sleepless condition, to administer the ammoniated tincture of lupulin, made by William Neergaard, the chemist of this city, in twenty-minim to one-drachm doses, or Fothergill's solution of hydrobromic acid in thirty-drop doses, in water, at bed-time. I have found the constant current of electricity useful in the form of centric galvanization. To recapitulate: Catalepsy comes on suddenly, generally after mental or emotional disturbances; the body becomes corpse-like and pale, the respiration being slow, and the pulse very soft and, perhaps, not discernible. The patient cannot be roused, and sensibility is lost. The stiffness of the muscles is a diagnostic feature of the disease, which is such that if a limb be put forcibly into any position it retains it. Patients remember nothing of any attack or what transpires during its continuance.

In all states of unconsciousness, where there is disseverment of the will from the organs habitually acted on by it, and during which odd, eccentric or dangerous acts are committed, it would seem most probable that while memory is annihilated, the acts are the outcome of the sensations, ideas, emotions, acts and events of antecedent life, and not inventions new to the senses. I doubt if the mind ever actually ceases in its operations or workings; and it is probable that actions analogous in kind, although variously altered in operation, occur in the brain, alike in unconscious and conscious states, in much the same manner as they occur in the sleeping and in the waking brain. It is a very difficult matter to try to define or explain mental action in these states, because there are as many forms and degrees of disordered mental action in states of unconsciousness as of the intellectual and moral qualities in their sane state. The confused and perverted notions of right and wrong in opium habitués, where the opium dulls and deadens the moral sense without seeming to disturb the intellectual faculties — owing to the close relation between opium and consciousness — have a very interesting medico-legal bearing, as these cases inhabit, more or less of the time, a realm of partial unconsciousness ; but this subject is too complex to admit of further mention.

The instigation to give way to inexplicable and ungovernable impulse, to cry out or shriek, to perpetrate a homicide or suicide, or to commit some motiveless act of violence or otherwise, and some of the acts of kleptomaniacs, come under the head of states of unconsciousness.* In families where madness is hereditary, there would seem to be a similarity or identity of the inner nature of different members of the same family, which would appear to incite them to the act of self-destruction without any appreciable incentive to the act. The suicidal act or deed in such instances is probably committed during a temporary partial state of unconsciousness.

In the case of the young English lady of wealth and refinement, who, while expensively dressed, took a greasy piece of meat from a butcher's shop, placing it between her velvet jacket and her silk dress, and walked off with it — is it rational to suppose that she was conscious of what she was doing? There is certainly a modified state of consciousness in kleptomania which makes the victim of this unhappy disease but very imperfectly conscious of the nature of the act. When this morbid propensity

---

* At the time of writing this a patient, fifteen years of age, tells us that she should feel "so much better if she could only kill somebody." Who, she cares not. There is a strong impulse to commit the act. This patient subsequently endeavored to commit suicide at her residence by jumping from a window.

appears, it generally comes on suddenly, and is, I think,
owing to some peculiar change in the nervous constitution of the woman — for it is generally women who are
affected with this type of nervous disorder. An uncontrollable impulse seems to usurp the whole mind for the
time being, and efface all other impressions. It seems to
annihilate personality by excluding all the relations
which determine it. I have, in my own mind, determined the invasion of insanity in patients who acknowledged such instigations as I have spoken of, to homicide
or suicide, which they had not given way to, but which
had excited their amusement rather than their appreciation, as in a sound mind would be the result. A patient
of mine, who appreciated his own condition, confessed to
me that he dreaded to look at children, because, although
he was very fond of them, he felt irresistibly impelled to
kill them. He related to me a struggle in his own mind
which occurred upon seeing a child upon the deck of a
steamer, in which he successfully resisted the impulse to
throw the child overboard. He said that he experienced
a dreadful mental contest, and that his head swam and
that everything looked black before his eyes. He knew
perfectly well that it would be wrong to commit such an
act, but his will-power was very nearly overthrown by the
disease. A lady, who was under my care, was irresistibly impelled to suddenly shriek aloud at any moment,
and struggles hard against these impulses. She is accustomed to have momentary periods of insensibility —
caused, I think, by anæmia of the brain — in which she
steadies herself by a table or chair, and generally man-
ages to avoid the observation of those in the room. This
lady, although sane in the eyes of the world, has twice
attempted suicide, and, in common with other insane
acts, these attempts have never caused her a moment's
regret, although I have repeatedly endeavored to elicit
such an expression from her. There is a taint of mental
disorder in the family. Another lady lately came to this
city from Massachusetts, to consult me about an irresisti-

ble impulse to throw herself from any horse car, steam car, steamboat or moving vehicle she was in. She deeply deplored this impulse, but it completely overpowered her, and she lived in fear that she should give way to the impulse. She had a lady friend accompany her constantly. In this case the cause was evidently dependent upon anæmia of the brain and spinal cord. An appropriate course of treatment cured her. There was not the element of insanity in her case. She complained, however, of lack of complete consciousness at such times, which fact she appreciated, and therefore never trusted herself alone. I desire to call especial attention to the fact that there are, preceding many states of unconsciousness, premonitory conditions of sadness, peevishness, irritability, quarrelsomeness, torpidity of conception, failure of memory, obtuseness of ideas, hebetude and prostration, followed, as the climax appears, by excessive gaiety, excessive exaggeration of physical strength, restlessness, vertigo, and passionate outbursts of fury. There are also in these states headache, vomiting, and neuralgia.

These constitutional states of morbid action show us that it would be very difficult for the mind to act calmly or clearly, and they also show a predisposition to actual mental disorder. These premonitory symptoms should always be inquired after, in medico-legal investigations, as they are really a part of the diseased state of the nervous system, and often precede the outward explosion for months. They constitute a part of the disease in the same way that the premonitory aura constitutes an integral part of the epileptic fit, when it is present.

I think more importance should be attached to the subject of uncontrollable impulse, and the legal profession should believe in its existence. At present, acts of unconquerable and destructive impulse, occurring in persons whose sanity has never been disputed, are generally visited by the extreme penalty of the law. These persons, however, I think, suffer from a condition not unlike the first stage of epilepsy when the pallor of the face occurs. In these cases of uncontrollable impulse, there is a condition of vascular tonus causing pallor of the face before the act, and the impulse ceases upon the commission of the act. I contend that in many of these cases there is a disease of the brain, and that many of these persons are morally irresponsible, especially as it has been shown that these impulses are recurrent. The uncontrollable impulse is unlike epilepsy, in that there is no complete and sudden loss of consciousness, while it resembles it, in the recovery being rapid and in the fact of the patient having no remembrance of the attack in many cases. These patients will tell you that they feel an ungovernable impulse to "do something." If the "doing something" consisted of undressing and shrieking from the top of the house, you would all say, "Poor creature! she is insane;" while, on the contrary, if the same person seized a knife and committed a murder, the people would assuredly hang her, although the deed would be equally that of a temporarily insane woman, committed during a state of partial unconsciousness — for I hold that these individuals are only very imperfectly conscious of their deeds. I consider these attacks as closely analogous to incomplete and abortive epileptiform attacks, and this should be accepted, I think, as their medico-legal significance. In these incomplete epileptiform attacks there are no convulsions and no complete loss of consciousness, the period being a mental blank to the patient, or a gust of passion, or a slight incoherence, or a slight vertigo, perhaps. I think that there is a functional brain disturbance in these cases of uncontrollable impulse, consisting of disturbance of the vaso-motor nerves, which are distributed to the blood vessels of the brain and form their calibre, the disturbance consisting of a condition of spasm of the blood vessels and temporary anaemia of the brain, evinced by the pallor of the face, which, as I have said, accompanies the uncontrollable impulse and generally characterizes it.

The difference between the epileptic state and that of the brain in uncontrollable impulse is, that in the latter case the state of anæmia is not followed by the congestion and hyperæmia which in epilepsy immediately follows, as a rule, the state of anæmia. The motor tract of the brain and spinal cord is probably not affected as in epilepsy. If this uncontrollable impulse led to suicide, would you not consider it as the deed of a person who temporarily was of unsound mind? If so, should not the impulse leading to murder deserve any amenity and leniency in treatment? I think that uncontrollable impulse, in common with epilepsy, insanity, chorea, etc., has a common origin, that origin being constitutional disease, or hereditary disease, which has been transmitted from some member of the family, more or less remote, to the patient under observation. It is a medico-legal point of great importance, which should be borne in mind, that there is a correlation of morbific forces — first thoroughly demonstrated by Dr. J. M. Winn, of London, England — which applies to a large class of hereditary diseases, making them mutually convertible ; in other words, that there is, in hereditary disease, a latent morbific force, which accumulates, perhaps gathering intensity during the latent period, and finally manifesting itself outwardly by a maniacal attack, in the convulsive movements of epilepsy, in consumption, in a suicidal act, or in the giving way to an uncontrollable impulse to jump from a house, kill a child, or violently swear and use obscene language when the general moral character may have been for months most unexceptional. It is a terrific thunderstorm of the mental and moral nature, due to the explosion of this subtile morbific force, which may have remained latent for a long time. The point which I desire to impress is this : that if, in medico-legal investigations, the judiciary in all such cases will take the same trouble to institute close inquiry that an experienced physician does, they will, in many cases, easily discover the existence of hereditary disease, which greatly modifies the prisoner's moral and legal responsibility ; and surely every prisoner is entitled to the benefit of such an investigation, if it is claimed that the criminal action was the offspring of disease which was not under the control of the unhappy sufferer. I hope I shall so convince my readers that it may be said, in after years, that the medical profession is entitled to the credit of inaugurating in this country the reforms so much needed. I have been told once or twice by legal friends that these were dangerous doctrines ; but I hold that a scientific truth is never a dangerous doctrine, and I do not believe it is right ever to sacrifice a human life to a cautious conservatism that fears to accept a truth because that truth may be in opposition to traditional dogma. I come, finally, to the most important of the states of unconsciousness, that connected with the disease of epilepsy.

Epilepsy is a functional disease of the nervous centres, the phenomena of which morbid state consists in seizures, generally sudden in their invasion, and preceded, as a rule, by a well-marked prodromal period, characterized by a loss of consciousness, coming on suddenly, and attended by peculiar involuntary muscular movements, which are highly spasmodic and convulsive in nature. There is great medico-legal importance attaching to epilepsy, from the reason that *there have probably been more grave crimes committed by persons epileptically insane than during all other states of unconsciousness put together in the annals of medicine and law.* I will go farther and say that I believe most of the *revolting and motiveless* crimes in the annals of history to be due to the epileptic state. Revolting and motiveless crimes often form substitutes for the epileptic paroxysm, just as periods of faintness or automatism often take the place of a fit. Whether or not the tragedies, like the one I shall shortly relate, are ever premeditated in imagination during the period of incubation of the fits, is, I think, a very difficult question to answer. The state of unconsciousness occurring in epilepsy may be substituted by any grade of sudden acts of fury or violence, homicide or suicide. A premonition of an epileptic fit has been followed by a state of unconsciousness, during which, instead of having the convulsion, persons have walked long distances, in one instance as far as eight miles. The recollection in this instance was a complete blank. If any catastrophe had been the result of this period of walking coma in these cases, during which time there was a total suspension of present knowledge and memory, I am afraid that the plea of temporary unsoundness of reason would have been looked upon with decided suspicion; yet the series of psycho-physical disturbances in these cases, whether apart or identified with an epileptic diathesis, directly affects the soundness of mind. The most insidious of these states of unconsciousness is that which dates from the close of the *grand mal*, or fully developed epilepsy, with convulsions. This state may continue for some days after an epileptic convulsion, and the patient appears so much like himself as to deceive even his friends into the belief that he is mentally normal. This state seems to be compatible with many rational actions, and its existence is not generally suspected until the commission of some crime, like the poisoning about to be spoken of, which succeeded a nocturnal fit of epilepsy. My opinion is, as it will be seen in the narration of this case, that there should be immunity of punishment to epileptics for criminal acts committed within three days before or after an attack, such act being evidence to me of mental unsoundness.

In conclusion, I will speak of the psychological aspect of the Laros case, on the trial of Allen C. Laros, at Easton, Pennsylvania, for the murder of his father, Martin Laros, by poison, the defense being based upon the allegation of epileptic insanity. The history of this very interesting case was kindly given me by my friend, Henry W. Scott, Esq., of the Pennsylvania bar, to whom I am indebted for it.

The Laros family lived at Mineral Spring, situated on
the Delaware river, in Northampton county, four miles
above Easton, Pennsylvania. The little hamlet consists
of a tavern and the homes of seven or eight families near
together, along the river road. Martin Laros, the father
of the family, was fifty-seven years old, and his wife
was fifty-one. They had lived at Mineral Spring for
thirty years. He taught school during the winter
months, worked his farm in the summer, and at the
same time was employed as undertaker and cabinet-
maker. He was quiet, unobtrusive, and respected in his
neighborhood. Mrs. Laros was a woman of domestic
habits and lively temperament. They have had seventeen children, thirteen of whom are now living. Several
of them have been school teachers. Some are living in
the neighborhood, and others have removed to a distance. At the time of the poisoning the family consisted of the father and mother, Allen (the prisoner),
Erwin, Alvin, Clara, Alice, and a very young grand-
child. Moses Schug, also a member of the household,
was a bachelor, sixty-two years old. He assisted Martin
Laros on the farm and in the shop.

One evening, while the family were at the supper-table,
they were, one by one, taken violently ill. Neighbors
came in to do what they could for the sick, and physicians were summoned. Allen also assisted in caring for
the sick ; he was taken sick later in the evening. Mrs.
Laros died at seven o'clock the next morning. Mr.
Laros also died on the same day, about noon, and Moses
Schug at three o'clock on the following afternoon. The
other members of the family recovered in about a week.
The fatal supper was partaken of on Wednesday. The
coroner's inquest was begun on Thursday afternoon, and
on Saturday the following verdict was rendered : "That
the said Martin Laros, Mary Ann Laros, and Moses
Schug came to their deaths from the effects of arsenic
poison, administered in coffee, on Wednesday evening,

May 31st, 1876, and that we believe the same was administered by Allen C. Laros."

A warrant was issued at once. Young Laros was arrested as he lay sick in bed, and taken to the county prison at Easton, Pennsylvania. The prisoner was about twenty-six years of age, a little under the medium height, and slightly built. He had received an ordinary common-school education, and was fairly intelligent. He was temperate, industrious, and moral, and was a church member. He was always disposed to be somewhat reticent, and spent much of his time alone. He was of respectable parentage, of healthful surroundings, of good moral and intellectual training, a teacher of the young in one of the public schools in his own township. He was, however, an epileptic, the epilepsy manifesting itself more than four years before the poisoning took place, and had continued, by successive steps of longer or shorter duration, until the time of the poisoning. For three weeks before this time, almost daily, he was so afflicted with epileptic convulsions as — so counsel for defense claimed — to dethrone his reason and destroy the powers of the mind. It was claimed and proved that, on the Saturday previous to the crime, he was afflicted with convulsions; that he had them on Sunday, Monday, Tuesday (the day the Commonwealth claimed he bought the poison), on Wednesday (the day of the poisoning), and on Thursday and Friday, immediately after it. After his confinement in prison he was similarly affected by these convulsions, varying in duration from a few minutes to several hours. During the continuance of the convulsions he was totally unconscious. Before and since his confinement, for a period of several hours after these convulsions had passed away, his mind was cloudy and confused, and his conversation and acts not responsible.

My own opinion has always been that, in the event of a criminal act by an epileptic, we should suspect mental disorder, and that, in the absence of any strong personal motive, there should be immunity of punishment to epileptics for acts committed within three days before or after an attack, such insane acts being to me the evidence of an insane mind. Such persons are, I think, able to conduct their business, and perform their duties, and continue their pursuits in all respects like other people, except at the time of seizure. In the case of young Laros there was an inherited tendency to insanity and nervous diseases for several generations, and in many branches of the family of the prisoner — grandfather, grandmother, and maternal aunt. These circumstances all contributed to lower the grade of his offence, even if it was not the offspring of decided insanity.

While young Laros was in prison awaiting trial every possible experiment was tried to ascertain if he were conscious while in the convulsion, and every conceivable test applied to see if the prisoner were feigning. The prison physician, during the first paroxysm he witnessed, suddenly thrust the blade of a sharp knife into the prisoner's hand, and no sensation was manifested. A heated key was next applied. Then the flame of a lighted lamp was held to the sole of his bare foot, and still not a quiver of sensation followed. Melted sealing-wax was dropped upon the bare skin so that the sealing-wax burned into the skin, and no indication of pain was shown. Nothing that science could suggest was left untried to detect imposture, if any existed, but all these tests failed to detect any feigning on the part of the prisoner. At the trial, Dr. John M. Junkin, of Easton, Pennsylvania, testified that he was called upon to visit Martin Laros on the morning of June 1st. Reached there about three o'clock, and, concluding from the symptoms that they were all suffering from arsenical poison, he gave stimulants and hydrated peroxide of iron. He found his patients vomiting and purging, and gave it as his opinion that the death of Martin Laros was caused by arsenic. During the progress of the trial various persons testified to having been aware of the prisoner's infirmity, and the deputy warden of the county prison testified as to the nature of the attacks while Laros was in prison. He described finding the prisoner "struggling in his cell in a fit," with his face very white, eyes partly closed, the hands clenched, with the thumbs inside, and that he heard the prisoner's teeth gritting. He also described incoherent and apparently insane conversation of the prisoner, and hallucinations of sight. The prison physician also testified that he found him — with a weak and feeble pulse and cool, pale skin — acting in a wild, incoherent manner; talking about fishing, seeing water-snakes, and other nonsensical, insane conversation. Any bright object he would endeavor to get hold of. His pockets were stuffed with bits of paper and such things. He tried to get the warden's shoe-buckles and the bright tips of the doctor's shoestrings. The doctor also testified that he, the prisoner, did not appear to have good control over his muscular movements. The doctor also described various epileptic convulsions which he witnessed, and testified as to the total unconsciousness of the prisoner during the paroxysms. He also testified to seeing the prisoner six to eight hours before an attack, when he appeared dull, and gave imperfect answers, and complained of pain in the head. The prisoner's condition while under observation, coupled with the testimony of his friends as to his previous symptoms and condition, led all unprejudiced observers to believe that he was mentally unsound. Dr. John Curwen, the Superintendent of the Pennsylvania State Lunatic Asylum, testified that he considered frothing, swelled veins in the neck, and lividity of face as essential symptoms, and without these he would doubt the genuineness of the epilepsy, although, on re-examination by counsel for defense, he admitted that these signs might possibly be absent in cases even of pure epilepsy. Dr. Curwen was expert for the Commonwealth of Pennsylvania. The jury in the case rendered a verdict of

murder in the first degree, and the prisoner was duly
sentenced to be hung.

The death-warrant was signed, but a writ of error was
sued out in the Supreme Court of Pennsylvania, which
operated as a *supersedeas*, and the governor recalled the
warrant. The counsel for the defense then presented to
the court a petition alleging mental unsoundness, and
asked for a commission to inquire into the matter and
ascertain whether the prisoner was a proper subject for
capital punishment. The commission appointed by the
court consisted of Dr. William Pepper of Philadelphia,
Dr. S. Preston Jones, also of that city, associated with
Dr. Kirkbride, at his asylum, and Hon. Henry A. Ross,
a lawyer of Pennsylvania.

The commission spent a month or more in taking testimony and making a personal examination of the
prisoner. They made a unanimous report to the court
that he was an epileptic and mentally irresponsible, that
he should not be visited with capital punishment, and
recommended his removal to an asylum. Thereupon
the court ordered him to be removed to the State Lunatic Asylum, at Harrisburg, Pennsylvania, of which Dr.
Curwen is superintendent. After confinement for a
period of about two years he escaped, and subsequently
was captured in Arkansas, or, rather, he surrendered
himself to the authorities and requested them to "send
him back to this country to be hung." He didn't want
to be returned to the asylum. He was returned to the
asylum, and about six months ago he escaped from
there a second time, and nothing is now known of his
whereabouts.

The able efforts in his behalf and in the cause of
humanity are owing to the exertions of his counsel,
Henry W. Scott, Esq., of Easton, Pennsylvania. Upon
his examination the prisoner declared that his father
and mother were both living and that his father was
making a door when he left home. One of the prisoner's
brothers was, up to the time of his death, a quiet,

uncommunicative, and retiring man, and he died by
hanging himself without apparent motive or cause.
Young Laros was a person of uniformly mild and
tractable disposition, who was brought up amid the
softening and restraining influences of a pious and
affectionate family and away from demoralizing surrounding or vicious companions. This outrageous and
enormous crime was very likely the outcome of
mental disorder which had depraved and eclipsed
the moral faculties. Yet the judge and jury deliberately arrived at a verdict which doomed this unhappy
creature to the scaffold. In reviewing this case psycho-
logically we have, as I have said, a mild-mannered boy,
of previous exemplary behavior, uniformly kind and
affectionate, suddenly developed into an inhuman mons-
ter of depravity. For four years he had been afflicted
with epilepsy, and we must bear in mind the tendency of
epilepsy to generate the insane impulse to crime. We
must also bear in mind that there are on record many
homicides committed by epileptically insane persons
under every circumstance of apparent motive and design.
There was a rapid succession of the spasms shortly before
and after the Wednesday night on which the family were
taken sick. These attacks had been noticed more particularly during the few months preceding the tragedy, and
they had occurred with startling distinctness and frequency, and on the very evening of the murder he was
unquestionably under the influence which precedes and
follows the epileptic paroxysm of epileptic insanity.
The experts for the Commonwealth in this case adopted
the typical case of epilepsy as the unvarying standard by
which the disease is to be ascertained, and it was only
under the most rigid cross-examination that they would
modify, in some degree, this position. The symptoms
of epilepsy are not, however, invariable. There may be
every variety, from the simply vertiginous to the most
demonstrative muscular and nervous spasms. The epileptic may be pallid or purple-hued, the pupils may con-

tract or dilate, the fingers may be clenched or extended, there may be foaming at the mouth or it may be absent. That some of the symptoms of the most decided and impressive type are not present is no proof that the disease is not epilepsy. The disorder of the intellect which accompanies epilepsy is similar to that we meet with in chronic insanity, and while, of course, it is not the invariable rule, yet in my own practice I have, in the great majority of cases, observed enfeeblement of memory and intellectual powers amounting to insanity. While an epileptic may be very intelligent, I do not believe that, either during the attack or for an indefinite period subsequently, the mental faculties are under the control of the patient. The patient, particularly as the effect of the lighter seizures, becomes very irritable indeed, and there are instinctive impulses, I think, to acts of violence. The confused recollection of what has happened and the unconsciousness of the gravity of his acts is, I think, diagnostic of the mental state of the epileptic, and should be considered as the essential characteristic of it. The epileptic, in the majority of cases, seems to automatically obey the impulses generated by his disease, and seems utterly powerless to resist them, even though they impel to criminal deeds. This constant disturbance of the affective and intellectual faculties which is manifest after the paroxysm, may last during the greater part of the interval between the fits, and this is a medico-legal point of great importance. There may be abortive epileptiform attacks, where there are no convulsions and where there is no complete loss of consciousness,—a sort of epileptic vertigo,—and yet such persons have committed sudden deeds of violence and were utterly unable to remember the circumstances afterwards.

I think there are cases where the *petit mal* of epilepsy may continue for hours, where no overt act happens, and where there is no motive for falsification. There is also, dating from the close of the *grand mal*, an insidious and obscure state resembling healthy mentalization, and differing from it only by a complete unconsciousness compatible with many rational doings. This state follows the convulsion, and is very dangerous to those around the patient. The acts in this state are closely allied to the state of unconsciousness in somnambulism. There is no knowledge or recollection of events that occur, or of overt acts that may be committed during this state. Baillarger relates the case of a vine-dresser near Lyons, France, who was seized with a fit of shivering, and who took up a mattock and killed three of his children, and not but a few rods from that spot he killed his wife and last child. He was much attached to his wife and children. Falret relates a case of a youth liable to vertiginous seizure, so severe as to occasion him to grasp the nearest object for support, who attempted to poison himself, was not excitable, would leave his business abruptly, walk seventy-five miles from Paris, taking no food for forty-eight hours, would forget his ordinary work, would walk during the night, wounded a lady in the street and remembered nothing of the assault. The unconsciousness or mental weakness the sequela of epilepsy permits of the existence of delusions — morbid mono-ideaism, irresistible impulse, and murderous instincts, which regulate automatically the volition and acts of the patient. This is a scientific and well-attested fact. There is only partial responsibility in this state.

*The Mental Condition in Hypnotism.** Dr. D. Hack Tuke, in his address on this subject before the Medico-Psychological Association in London, February 21st, 1883, said that he had tried to form a clear idea as to the cerebro-mental condition of hypnotized persons. The data upon which we have to form an opinion or construct a theory are:

---

* I would define *hypnotism* as a morbidly profound sleep of the cortex of the brain while the basal ganglia remain unaffected and in their normal condition.

1st. The condition necessary to induce the state in question.

2nd. The objective symptoms of the hypnotized so far as we can observe them ; and,

3rd. The subjective state experienced and described by himself (the hypnotized person), in those instances in which memory, more or less distinct. is retained of what has been present to the mind during the hypnotic condition.

1. *As to the Condition Necessary to Induce the Hypnotic State.* Staring at a disk or some well-defined object is a very frequent method. Other methods are also effective : the monotonous sensory impressions produced by passes, by counting up to several hundred figures, by listening to the ticking of a watch, etc. We may throw ourselves into an hypnotic state in attempting to go to sleep. The *principle common to the various modes of hypnotism is on the physical side, the stimulation, more or less prolonged, of a sensory nerve in close relation to the brain, calculated to ultimately exhaust some portion of that organ, and on the mental side, the riveting the attention on one idea.* Looking at an object is not essential, for a blind person may be hypnotized, and in susceptible persons the merely expecting to be hypnotized is sufficient to induce it, the expectation in this case involving the concentration of the attention to one point.

Mr. W. North, Lecturer on Physiology at Westminster Hospital, thus describes his own feelings while hypnotized : "I have not the smallest doubt, that at first I succeeded in abstracting myself, as it were. from surrounding circumstances. I had been reading very hard for days past on the subject of intestinal digestion in relation to the bacteria produced, and I pictured to myself the interior of the intestine and its contents ; then I tried to picture a special form of bacteria, and while I was engaged in contemplating its changes of form I seemed to lose all consciousness of persons around me."

On a subsequent trial being made, he looked at his boot, and thus described the process: "I ultimately succeeded in fixing my attention on six points of light reflected upon my boot, and having some minute resemblance in position to the constellation Orion. After looking fixedly at this for what seemed to me a very long time, the idea of the constellation vanished, and its place was taken by the outline of the lower part of the face of a friend. All I could see was his beard and mouth and part of his nose and one cheek, the rest was cut off by a broad black area; the details were tolerably vivid."

The voluntary surrender of the will — the subject placing himself passively in the hands of the operator, is also an important factor in nearly all the processes. It is the initial step to the subsequent abandonment of the will of the subject to that of another. M. Richet, of the Salpêtriére, has shown that the subject may be surprised, and even rendered cataleptic, the moment his attention is in the least arrested. He is seized, and, as it were, instantaneously petrified, whatever efforts he makes to resist the influence. M. Richet constantly produces hypnotism by throwing a brilliant electric light upon the face of persons not expecting it, or by striking a gong which had been concealed. An hysterical or neurotic subject has been transformed into a statue by a blow or the concealed gong at the Salpêtriére.

2. *The Objective Symptoms of the Hypnotized.* These vary with the stage or type. Charcot, Richet, Tamburini and Sepelli recognize three fundamental types, the cataleptic, the lethargic, and the somnambulistic. In the first the limbs retain the positions in which they were placed, for a considerable time and without effort; in the second (the lethargic), the muscles which are relaxed are found to have the remarkable property of contracting in a most definite way under gentle mechanical application; in the third (the somnambulistic), the state of the subject answers much more to what is understood, as the so-called magnetic or mesmeric sleep. Contraction of

the limbs can be produced, but they are of a different
character from those in the cataleptic form, or the excitability of the muscles in the lethargic state.

Pupils. The pupils exhibit strabismus and contraction,
and afterwards are widely dilated and sluggish, an indication of the functional activity of the medulla, as regards
the sympathetic as well as the respiratory centre.

Cerebral Circulation. Ophthalmoscopic examination
by Professor Förster of Heidenhain's patient showed that
there was no contraction of the vessels as Heidenhain
expected to find, as his theory had been that anæmia
caused the sleep. That hyperæmia of the brain is not
inconsistent with hypnotism was proved by hypnotizing
a gentleman (Heidenhain's brother), who had inhaled
nitrite of amyl. The respiration and pulsation are
quickened at first. Professor Tamburini used the pneumograph, and he found the frequency of respiration to
be doubled at first, and the inspiratory pause suppressed.
These tracings are useful in detecting simulation. With
the cataleptic subject the tracing is uniform in character
from beginning to end. With the simulator, on the contrary, it is composed of two distinct parts. At the
beginning, respiration is regular and normal : in the
second stage, that which corresponds to the indications
of muscular fatigue, irregularity in the rhythm occurs
with deep and rapid depressions, manifest indications of
the disturbance of the respiration caused by the effort to
simulate. Professor Tamburini made careful pulse
tracings also. The rise in the pulse is 100 per cent.
The myograph, the pneumograph, and the sphygmo-
graph are most valuable means, placed at our disposal
by modern invention, for obtaining trustworthy records
of the objective symptoms of hypnotism. There is
heightened reflex action. The tendon reflexes may be
normal or exaggerated. Richer states that in the lethargic type they are much exaggerated, in the cataleptic
type they are diminished, and in the somnambulistic
type normal. There is galvanic reaction.

3. *Subjective Symptoms Described.* Sensation of pain is deadened or suspended. Anæsthesia is produced. Mr. North said that a pin plunged into his hand nearly up to its head, felt as if a match or some blunt instrument were pressing against the hand. When he was roused it hurt him considerably to withdraw the pin. The special senses are interfered with or abolished. They may be either heightened or abolished in different cases. *Sight* is partially affected. The subject sees, though confusedly, that which is immediately around him, but has a very vague or no perception at all of what is beyond this range. Some subjects describe a play of colors before the eyes. *Hearing* is not affected. *Taste* is suspended. There may be no unconsciousness whatever in some instances, and the subject may appear like other people. A certain susceptibility to impressions on the mental side and to rigidity of the limbs on the physical side may be all that marks the state of the subject. Is it that the cerebral cortex is just sufficiently weakened in function to have lost its supremacy, without parting with its more secondary offices?

Volition. There is no spontaneity in hypnotized persons. Volition is suspended.

Extreme Susceptibility to Outside Suggestions. The subject hypnotized is without any will-power, and at the mercy of any suggestions, however absurd. Hallucinations are easily induced. A person may eat heartily while hypnotized, and their visceral sensations will not suffice to inform them, so that they will wish for the next regular meal as if they had not eaten. Richet, of France, says: "The somnambulist has a perfect memory, a very lively intelligence, and an imagination which constructs the most complex hallucination." The great fact in mesmeric sleep is that will and consciousness are suspended, and the brain placed in the condition of the true spinal or reflex system. There is a reduction to a mere automatic condition. Heidenhain hold that the cause of the phenomena of hypnotism lies in the inhibition of the activity of the ganglion cells of the cerebral cortex by prolonged stimulation of the sensory nerves of the face, or the auditory or optic nerve. A sensory nerve may certainly inhibit the brain centres, and this inhibition is the starting point of hypnotism.

Conclusions. 1. There may be consciousness during the state of hypnotism, and it may pass rapidly or slowly into complete unconsciousness as in the somnambulistic state; the manifestations not being dependent upon the presence or absence of consciousness, which is merely an epiphenomenon.

2. Voluntary control over thought and action is suspended.

3. The reflex action, therefore, of the cerebral cortex to suggestions from without, so long as any channel of communication is open, comes in play.

4. While the consciousness is retained, the perception of the reflex or automatic cerebral action conveys the impression that there are two egos.

5. Some of the mental functions, as memory, may be exalted, and there may be vivid hallucinations and delusions.

6. Unconscious reflex mimicry may be the only mental phenomena present, the subject copying minutely everything said or done by the person with whom he is *en rapport.*

7. Impressions from without may be blocked at different points in the encephalon, according to the areas affected and the completeness with which they are hypnotized; thus, an impression or suggestion may take the round of the basal ganglia only, or may pass to the cortex, and. having reached the cortex, may excite ideation and reflex muscular actions, with or without consciousness, and wholly independent of the will.

8. There may be in different states of hypnotism exaltation or depression of sensation, and the special senses. There is a peculiar abnormal mental condition presented in hypnotism, closely allied to mental disease, and full of interest to students of mental science. The subject has been scientifically studied by James Braid, of Manchester, in 1843; Esdaile, in India, in 1846; Girard Teulon and Demarquay, in 1860; Richet, in 1875; Charcot, in 1878; and, in or about 1880, by the late Dr. George M. Beard, Drs. Weinhold, Preyer, Berger, Grützner, and Heidenhain, and Dr. H. Charlton Bastian. We may fail at first with a subject, and after a few trials he may make an excellent subject for experimentation. Bastian says that persons, who have been once hypnotized, can in general be again brought with comparative ease into the same condition, and the facility of hypnotizing such persons goes on increasing after each operation, owing to the existence of a predisposing mental state. A condition of excited expectancy is a decidedly favoring mental state.

The simplest condition necessary to induce the hypnotic or trance-like condition is to make the subject look fixedly for a few seconds at a bright object, held by the operator at about eight to fifteen inches above the eyes, at such a distance above the forhead as may be necessary to produce the greatest possible strain upon the eyes and eyelids, and enable the patient to maintain a steady, fixed stare at the object. We must tell the subject to keep his eyes steadily fixed on this object and his mind riveted upon the image of it. In some persons, after fifteen or twenty seconds, we shall find a decided cataleptic state induced, so that the limbs have the tendency to remain in the position in which we place them, and, if not, we may gently request the patient to keep his limbs in the position in which we have placed them. The pulse now quickens and the limbs shortly become rigid. By prolonging this process we induce a profound sleep, or trance, in which there is complete anæsthesia.   Esdaile, in India, performed numerous operations on Hindoos with absense of all pain while hypnotized.   The therapeutic value of hypnotism has never yet been thoroughly tested, and the future may develop facts of much interest and value.

CHAPTER X.

Medico-Legal Relations of Inebriates.

Society in general to-day is more willing than formerly to accept the conclusions of science respecting the disease of inebriety. This is due to the more intelligent attention given to inebriety by means of institutions devoted expressly to its treatment that have been established in our own country and abroad. We have no work in which the various forms and degrees of inebriety are treated in reference to their effect on the legal relations of man; no work entirely devoted to the legal relations of inebriates. We need a complete and methodical treatise on inebriety in connection with its legal relations, in which the subject shall be treated in a spirit corresponding to the present condition of the science of inebriety.* The principles of law which have been laid down regulating the legal relations of the inebriate were framed long before we had obtained any accurate ideas respecting the disease of inebriety, and therefore great injustice has been done to the subjects of this disease under the name of law. Instead of kindness and consideration and good medical care and treatment and efficient nursing, the inebriate has received loathing

---

* Dipsomania is a disease characterized by a condition of mind in which the person is involuntarily deprived of the consciousness of the true nature of his acts, and by an irresistible impulse to drink alcoholic liquors, rarely stopping short of complete intoxication. There may be intervals of sobriety of *hours, days, weeks* or *months*, but the craving for alcohol, when it comes in, cannot be controlled by the patient.

and ostracism at the hands of his friends and acquaintances. We would premise our further remarks by laying down a general proposition which we hope may prove the corner stone of a medical jurisprudence of inebriety, viz. : *That the disease of inebriety should be regarded as exempting from the punishment of crime, and, under some circumstances at least, as vitiating the civil acts of those who are affected with it.* The difficulty in determining who are really the subjects of disease must be met by drawing a sharp line between the various forms of drunkenness and the *disease*, with its essential psychic and physical signs; between the individual who apparently chooses to indulge in alcohol, and him who is irresistibly impelled by the craving — often periodical — resulting from a morbid irritation of the cortical sensory centres of the brain, to indulge in alcoholic stimulants and to frequent fits of intoxication. As against the crude and imperfect notions that even high legal authorities have entertained of the pathological character of the disease of inebriety, we would place the results of more extensive and better conducted inquiries, the offspring of the steady advancement of medical science. The day, we think, is gone by when the accumulated results of experience in this department of science can be successfully contradicted by men utterly destitute of any knowledge of the subject, on which they tender their opinions with arrogant confidence; and the day is not far distant, we trust, when such men only shall be considered capable of giving opinions in judicial proceedings relative to inebriety as are physicians eminent for their knowledge of the disease of inebriety, and who have particular knowledge and skill relative to this particular disease. The single fact of the presence of mental disease should be sufficient to annul criminal responsibility, and dipsomania is eminently a mental disease. We think we fairly state the known facts of science, and the current facts respecting the disease of inebriety, when we say that clinical investigation of facts reveals generally an inherited neuropathic condition, an abnormal state of the nutrition and circulation of the brain and nerve centres, great irritability of the cerebral cognizant centres, morbid fears and dreads, morbidly-colored perceptions, conceptions, and misconceptions, timidity, irresolution and irritability of manner and speech, all of which are foreign to a healthy person ; all these are the physical characteristics of the neurasthenic stage of inebriety. We have here all the signs and symptoms of an abnormal condition of the centric nervous system demanding stimulants which constitutes the disease — inebriety. We have here a morbid psychosis, a disease of certain parts of the brain, resulting from some morbid irritation of the cortical sensory centres of the brain or from special molecular changes in the centres, perverting brain function, a condition markedly hereditary, and evinced outwardly by great nervous irritability or restlessness, unnatural sensations, an uncontrollable desire for alcoholic stimuli, and a disposition to frequent fits of intoxication.

There is a departure from a healthy structure of the nervous apparatus as in mental diseases generally. The inebriate is simply the subject of a disease, in which normal function is acting under abnormal conditions, and we should recognize this fact both as to medical and moral treatment and in reference to the legal relations of the inebriate. The pathological evidence in favor of these facts which I have stated, was at first slender, has been yearly increasing, and is to-day conclusive and unanswerable. There is a modified mental responsibility in this disease as in other forms of mental disease, and the common law should be codified to recognize the teachings of science. The code should contain a provision like the following, respecting the disease of inebriety.

By reason of their impaired responsibility, punishment cannot be inflicted on those who commit penal acts in a state induced by the disease of inebriety, which either takes away all consciousness respecting the act generally and its relations to penal law, or in conjunction with some peculiar bodily condition, irresistibly impels the subject of this disease, while partially or completely unconscious, to violent acts. Responsibility should be annulled in that condition, in which either a consciousness of the criminality of the offence, or the free will of the offender, is taken away by disease. If we say that the disease of inebriety is a form of mental disease, and that an act done by a person in a state of mental disease, or any condition of mind in which the person is involuntarily deprived of the consciousness of the true nature of his acts, can be punished as an offence, we then protect the inebriate satisfactorily. Please remember that by the very nature of the disease (the great diagnostic mark of which is the irresistible *impulse* and *craving* for stimulants), the person is involuntarily deprived of the consciousness of the true nature of his acts. In this case, the will is overborne by the force of the disease. The man's free will is taken away from him by the superior force derived from disease exactly as in the periodical insanities. There is no truer periodical insanity than dipsomania. The reflective and perceptive powers of the mind are markedly affected by this disease. The mind in dipsomania has no power to examine the data presented to it by the senses and therefrom to deduce correct judgments, neither can it perceive and embrace these data. The mind does not possess its ordinary soundness and vigor, and the existence of delirium at any period of this disease would seem to throw some degree of suspicion on any contracts entered into during such disease, and on the testamentary capacity of the mind. We should, however, be decided regarding this point by the circumstances that attend the making of a will, the previous intentions of the testator and the nature of the case. The testamentary capacity, therefore, of an inebriate is to be determined, in a great measure, by the nature of the act itself.

Whether inebriety should be considered a valid reason for divorce when concealed from one of the parties previous to the marriage, we would say, that in our opinion each individual case ought to be decided solely on its own merits.   We are inclined to place the inebriate on the same footing with one who labors under hallucinations.   He does not enjoy the free and rational exercise of his understanding, and he is more or less unconscious of his outward relations ; ergo, none of his acts during the paroxysms can rightfully be imputed to him as crimes.   The acts of an inebriate certainly proceed from a mind not in the full possession of its powers and oftentimes excited by unfounded delusions, and an enlightened sense of justice revolts from ever regarding them in a criminal light.

With reference to the suicide of inebriates we think that their views of persons and things are greatly confused and distorted, and that such persons are in such a degree of perturbation that they are unfitted for mature, correct, judgments ; and that if their suicidal designs were in any given case to be frustrated and the patient cured, it is not at all unlikely that we might hear the declaration that such a one was entirely unconscious of having attempted such an act.   Suicides and homicides, by those affected with the disease of inebriety, are done in a dream-like state of partial unconsciousness, in which the patients rarely know what they are about.   There is a very doubtful mental condition at the moment of the act, so that a jury are amply justified in acquitting such a person as "not guilty, on the ground of insanity."   I grant that it is often a difficult task to determine exactly the mental condition of an inebriate at the moment of his committing a criminal act, but I am inclined to believe that nearly always such a person is deprived of his moral liberty.   With respect to dipsomania, which term I would restrict to periodical attacks of inebriety, we have a true periodical insanity, characterized generally by excitement or depression and the irresistible craving for stimulants, which craving is allayed only by complete and deep intoxication. Succeeding this paroxysm of drinking is an interval during which the patient is rational and lucid, although there may be transient excitement during this lucid interval. In these intervals the dipsomaniac is as capable of transacting business as a person ever is in a lucid interval. There is a complete intermission of the disease, and this may last for weeks or months, but I think there is a weakness and irritability induced in the mind by numerous and frequent attacks or paroxysms, which unfit it for extraordinary efforts even during the lucid interval. Self-control is more easily lost, and there is a want of capacity for new or sustained mental efforts or responsibility felt by the patient himself. We do not think that these cases of periodical drinking or dipsomania, are, during their lucid intervals, either completely responsible or completely irresponsible for their civil or criminal acts. The mind cannot be affected of course, except through the brain, but as I have repeatedly before various societies detailed the pathological changes in the brain induced by alcohol, I shall not allude to them in this chapter further than to remark that of course such degenerative changes are directly related to the manifestations of the moral and intellectual powers of the subject of the disease of inebriety. The late eminent Dr. Ray graphically described the course of this disease, years ago, in these words: " With a full knowledge of the dreadful consequences to fortune, character, and family, he plunges on in his mad career, deploring, it may be, with unutterable agony of spirit, the resistless impulse by which he is mastered." It is, I think, a fact not generally known, that Esquirol distinctly recognized the disease of inebriety in its continued and periodical form, and termed it *dipsomania*, and attributed it to the influence of pathological changes, and absolved its victims as not morally responsible. (See note in Hoffbauer, § 195, and *Maladies Mentales*, 11, 80.) Esquirol says, " this

craving is imperious and irresistible," that "dipsomaniacs obey an impulse which they know not the
power of resisting," that they are "true monomaniacs."
He also says — and I invite the attention of the
legal profession to this emphatic statement of the
founder of psychological medicine — that we shall
find in these cases "*all the characteristic features
of partial madness.*" Esquirol relates the case of a
merchant about forty years of age who became gloomy
and disquieted respecting business reverses, neglected his
business, became irritable and ill-tempered. His tastes
and habits changed ; he commenced a course of inebriety, and neither the dictates of affection nor the authority
of his father availed anything. This was during the
winter. At the approach of spring the drink craving
ceased. He resumed his regular and sober habits, and
applied himself to business and showed a return of
affection toward his family. In the following autumn
appeared the same phenomena and the same spontaneous
cure in the spring. During the two following years the
disease ran its course, with its paroxysms and intermissions, until Esquirol finally cured him. The same distinguished authority relates the case of a lady who, after
being melancholy for about six weeks, which condition
is generally the antecedent stage of active insanity, with
weakness of the stomach and indisposition to take the
least exercise, was suddenly seized with the strongest
craving for spirituous drinks, together with sleeplessness,
agitation, and perversion of the affections (these latter
being peculiarly distinctive of mental disease). For six
years, Esquirol says, these symptoms made their appearance annually, and continued two months, the perfect
analogue of other periodical insanities. Marc, another
celebrated authority on mental diseases, in "De la folie,
etc.," II, 605, says that dipsomania sometimes occurs in
women at the turn of life, as it is called, as a result of
the important changes which at that period take place in
the female constitution. He has met with many examples
16

of it in women who previously had exhibited all the
virtues of their sex, and especially temperance.   After
this affirmation and description of the disease of inebriety,
by these celebrated men of profound study and extensive
observation, together with the authoritative utterances
of such men as our late Dr. Ray ; Sir Thomas Watson, J.
Milner Fothergill, Dr. B. W. Richardson, Dr. J. Crichton Browne, Dr. Alex. Peddie, Dr. Francis E. Anstie,
the late Dr. Forbes Winslow, Dr. A. Mitchell, and Dr.
Norman Kerr, of England ; the late Dr. David Skae, of
the Royal Edinburgh Asylum, and Dr. David Brodie,
of Edinburgh ; of Dr. Hagstrom, and Dr. Magnus Huss,
of Sweden ; Dr. M. Magnan, physician to St. Ann Hospital, Paris, and Dr. Dujardin-Baumetz, and Audige, of
Paris ; and Dr. Krafft-Ebing, in Germany, who have all
been trying to bring about a coöperative public sentiment and legislation, and all of whom recognize dipsomania as a distinct form of mental disorder, it betrays,
it seems to me, the height of ignorance and presumption
to question the existence of such a disease.   It is greatly
to be regretted, that even in the nineteenth century there
is a most deplorable ignorance of the mental operations
of those afflicted with the disease of inebriety.   There is,
certainly, either a constant or a periodical morbid condition of intellect or loss of reason, coupled with an
incompetency of the person to manage his own affairs,
and this certainly should constitute unsoundness of
mind, in the legal sense.   In the Austrian code of 1803,
section 2, lib. c, inebriety "is made a ground of exculpa-
tion from responsibility, when not produced with a view
of committing the crime."   In the Prussian Landrecht,
page 11, title 20, section 22, it is intimated, that "a
criminal act, committed in a state of drunkenness which
originates in fault, is punishable for the fault only."
In the Bavarian code, article 121, "inculpable disorder
of the senses, or of the understanding," which includes
inebriety, is mentioned as one of the grounds that
exempt from responsibility.   "The Zurich project

of 1829," says Ray, "declares that one who commits a crime, in a state of inculpable drunkenness
of the highest degree, is punishable in the same
manner, as if he were under legal age." In the
present penal code of France, inebriety does not
absolve from the ordinary punishment of crime.
Their code is, like our own, very deficient on this subject, as they practically decide that inebriety, being a
voluntary and reprehensible state, can never constitute
a legal or moral excuse.

In England inebriety does not afford any relief from
the ordinary consequences of crime. In the disease of
inebriety, and in that form of it (*dipsomania*) which is
periodical, we desire to impress the fact that the act of
drinking cannot be called a voluntary act at all. It is
done in obedience to the *blind, irresistible* craving for
the alcoholic stimulus. It is, properly speaking, an involuntary act that unintentionally, and automatically oftentimes, leads to the commission of crime when such overt
acts are committed. To constitute crime, there must be
moral liberty and an intention to commit crime ; the dipsomaniac acts in obedience to a *vis a tergo*, derived from
a brain condition that he cannot resist. The dipsomaniac never willfully deprives himself of reason. We
wish to point out that by the present code of this State,
that even if dipsomania, or the whole disease of inebriety,
be admitted as a form of insanity, the points submitted
to the jury for their determination will be whether the
prisoner is capable of distinguishing between right and
wrong. If they conclude that he is, they will return a
verdict of "guilty," notwithstanding the fact known to
every man conversant with the insane mind, that three-
fourths of all the insane are perfectly able to distinguish
between right and wrong, and the fact that this test is
about a century behind the times as regards our present
knowledge of mental disease. This absurd legal test of
responsibility was introduced into the code of New York
with the confidence which ignorance of disease usually

inspires, by those who are evidently utterly unacquainted with the phenomena of insanity or of the actual operations of the insane mind. When will the law of the land cease to confound the unconsciousness of right and wrong with the powerlessness, through cerebral defect or disease, to do right? The insane and many inebriates act as the result of the morbid notions in the mind that spring from their disease, and they do not act in response to the suggestions of affection, of reason, and of common sense, as sane persons do. Finally, insanity, whose remote cause is habitual drunkenness, should be, we think, an excuse in a court of law for a homicide committed by the party while so insane, but not at the time intoxicated or under the influence of liquor. This is essentially the decision of Judge Story in the case of Drew, who killed his second-mate, Clark, while in a state of insanity with hallucinations, the remote cause of which was the excessive use of alcohol.

It has been my aim to adduce facts relative to the disease of inebriety that will tend to convince any unprejudiced person that the victims of this form of mental disease ought not to be held responsible for their criminal acts. I have no doubt that high legal authority would say that it would be a dangerous truth to recognize in estimating the degree of criminal responsibility. We would emphatically impress on the minds of everyone that a dipsomaniac is a true monomaniac, not morally responsible ; that the disease of inebriety proceeds from the same pathological causes as other mental diseases, and that, logically, if we grant a cessation of moral responsibility in the latter, we must do likewise in the former.

CHAPTER XI.

An Attempt to Codify the Common Law Relating to Inebriety, from a Clinical, Scientific and Forensic Standpoint.

THE teachings of science and the facts of observation respecting the study of morbid pscychology have led the author to consider the following propositions as applicable to and indicating the forms of inebriety which affect the legal responsibility of men, and indicating to whom should be allowed the immunities of inebriety : *

1. If the jury in any case believe, from all the evidence in the case, that a given act was committed by the prisoner while suffering from the disease of inebriety, they should acquit him on that ground.

2. To connect threats with an overt act, the jury must find that they were uttered maliciously, seriously, with the intent to execute them in accordance with the purpose expressed by the prisoner in a normal state of mind ; and that the overt act occurred in pursuance of those threats.

3. Under any circumstances the jury must find, in any

---

* We venture to make a medico-legal definition of dipsomania as follows, viz. : Dipsomania is a diseased state of mind, due to ill health, accompanied by more or less absence of self-control, and impairment in a marked degree of the intellect, the emotions or the will, and showing itself psychically, by depression, exaltation, or mental weakness, and by disorders of sensation, perception or conception ; also at times by abeyance of mental functions.

given case, that any threats and acts in question were the product of a normal mind.

4. If there is a lack of malicious purpose, of depravity of heart, and a diseased understanding or will, there is a lack of the essence of crime.

5. When there is a defect of the mind the will does not join with the act. If disease of the body affecting the mind deprives a person of a determination of his choice to do or to abstain from a particular act, there is loss of responsibility entailed on that person by reason of physical disease, viz., inebriety.

6. The primary part of any case of the prosecution where inebriety is alleged as a defense, to be established to the satisfaction of the jury beyond any reasonable doubt, is, that a person accused shall be possessed of an intact intellect, shall have shown evil designs, and that he has not been deprived by disease of the power of choice, or of moral liberty, and that the act was not the outcome of disease; and the normal state of the prisoner's mind is to be made out affirmatively by the prosecution as a part of their case.

7. Disease of the mind may deprive a man of understanding or of liberty of will, and may make him, therefore, incapable of appreciating the obligation of law, and should deprive him of his liability to punishment for violating laws which, by reason of diseased intellect, emotions, or will, he may not have the capacity to obey. In respect to a given act, such a man has not the normal consent of his will. Lord Hale says: "The consent of the will is that which renders human actions either commendable or culpable; as, where there is no law there is no transgression, so regularly, where there is no will to commit an offense there can be no transgression or just reason to incur the penalty or sanction of that law instituted for the punishment of crimes or offenses; and because the liberty or the choice of the will presupposed an act of the understanding, to know the thing or action chosen by the will, it follows that where there is a total

defect of the understanding, there is no free act of the will in the choice of things or actions." Therefore, no man can commit a crime *unless he has the control of his will.* An unwarrantable act without freedom of the will is no crime at all.

8. An individual should be exonerated from liability for his act, where, although he knew right from wrong, he was overborne by an impulse to drink he could not control, where he saw the wrong, perhaps, but had no power to abstain from the act.

9. In any given case, even if the evidence as to the inebriety of an individual should leave it in doubt as to whether he was intoxicated or not at the time of the commission of the alleged act, the individual is entitled to an acquittal if a dipsomaniac. If the jury entertain a reasonable doubt as to the perfect sanity of an individual at the time of the commission of such act, respecting such act, they are bound to acquit him.

10. If an individual, as the result of dipsomania, is temporarily thrown into a state of excitement, in which he is divested of his mental power to an extent placing him beyond the range of self-control, in reference to the particular act charged against him, so that he could not possibly restrain himself from the commission of the act alleged against him at the very time of its commission, he is entitled to an acquittal. Was he at the very time of the perpetration of the deed rendered by disease incapable of reasoning upon what he did, or of refraining from the commission of the deed ?

11. In psycho-sensory insanity, which is analogous to dipsomania, the affections — passions and emotions — are affected by disease, while the intellect may be, as far as we can discover, unimpaired. Hammond has correctly testified that the mental diseases to which these terms of moral or affective insanity are applied are real, undoubted diseases of the mind controlling human action. Bucknill and Tuke of England, in their Manual of Pscychological Medicine, so hold.

12. The knowledge of right and wrong with regard to any particular act is no test at all of a normal state of mind or mental responsibility. Fully one-half of the insane in asylums to-day perfectly appreciate this difference as well as sane persons do.

13. No act done by a person in a state of dipsomania can be punished as an offense, and no insane person can be tried, sentenced to any punishment, or punished for any crime or offense, while he continues in that state.

14. Dipsomania being thus established as an absolute bar to a criminal prosecution, which forbids responsibility for a crime or offense committed while in that state, it is immaterial how long it exists before or after the commission of the act. It is enough that it existed at the time of its commission. If mind is obliterated or diseased for *any* length of time, however short, it is a good reason for recognizing unaccountability.

15. If some controlling disease is the acting power within any individual, which he cannot resist, that individual is not responsible. We must remember that the moral as well as the intellectual faculties may be so disordered by disease as to deprive the mind of its controlling and directing power.

16. Judge Edmonds has said: "In order to constitute a crime a man must have memory and intelligence to know that the act he is about to commit is wrong, to remember and understand that if he commits the act he will be subject to punishment, and reason and will to enable him to compare and choose between the supposed advantage or gratification to be obtained from the criminal act, and the immunity from punishment which he will secure by abstaining from it. If, on the other hand, he have not intelligence and capacity enough to have a criminal intent and purpose, and if his moral or intellectual powers are so deficient that he has not sufficient will, conscience or controlling power, or if, through the overwhelming violence of mental disease, his intellectual power is for the time obliterated, he is not a responsible

moral agent, and is not punishable for criminal acts."
It will be seen that the learned judge had very correct
conceptions of mental pathology, as in the remarks he
includes intellectual insanity, moral or affective insanity,
and temporary insanity, and inebriety. He correctly
recognizes that the perception, emotions, and the will
may be affected, as distinct from the intellect. Also, that
the absence of the power of self-control, the result of
disease, is an essential element in responsibility.

17. There should be immunity resulting from a *recent*
or *sudden cause* which may deprive an individual of the
power of choice, of moral liberty, or of mental freedom
in regard to a given act.

In a given case, if an individual, at the very time of
the commission of an act alleged against him, from
causes operating for a considerable length of time beforehand, *or recently or suddenly occurring*, is mentally
unconscious of the nature of the act in which he is
engaged, he is legally irresponsible for it. The state of
mind at the time of an act is to be looked at in determining the character of such act. To put such a test to the
jury as the ability or capacity to distinguish between
right and wrong, is no standard as to whether a man is
or is not in a moral state of mind. Most of the insane
inebriates, when they commit crime, are carried away by
ungovernable feelings, and many of them can distinguish
between moral right and wrong. They are, nevertheless,
totally irresponsible, criminally, for their actions.
When the mind is in a state of frenzy a man is deprived
of his moral liberty, deprived by the uncontrollable
influence of all control over his will. *The intellect may
condemn an act which the will is powerless to restrain.*

18. We may have partial inebriety, in which either the
intellect, the emotions, or the will may be together or
separately affected by disease in a manner to completely
annul responsibility.

19. If we say that a person who has sufficient mental
capacity to distinguish between right and wrong, in

reference to its particular act, and to be conscious that it
is wrong, is in a normal state of mind, then we must
empty our insane and inebriate asylums of one-half of
their inmates, as they are by this test of normal mind
held there unjustly, in defiance of law.

20. If an individual, by reason of the disease of
inebriety, entertains false hopes, has delusions or
hallucinations of the special senses; has his natural
affections perverted by disease; cannot estimate correctly the amount of property he is possessed of; has
delusions of wealth or grandeur; has not the capacity
to exercise will in reference to his conduct, he has not
proper testamentary capacity.

21. *Testamentary Capacity.*   If an individual is intel-
lectually and morally sane in reference to the commission
or execution of a legal document; understands the
nature and amount of property he is disposing of; does
not ignore the claims of natural relationship; is not
influenced in the making of a will by delusions respecting those nearest and dearest to him, such an instrument is valid, even if the person making such an
instrument be the subject of mental disease or inebriety,
and confined in an institution for the treatment of mental
disorders, or in an inebriate asylum.

22. In delusional inebriety it must be remembered that
the premises reasoned from by the inebriate are uniformly
false. The intellect is diseased; the imagination is
diseased and disordered. Morbid delusions, and not real
circumstances, are the impelling motives to acts.

In the criminal inebriate a strict inquiry must be made
in relation to his former habits, disposition, and modes
of feeling and action. This will probably result in one
of two things: either a marked change will be found to
have occurred, which will be likely to date from the
period when he sustained some reverse of fortune, or
experienced the loss of some near and dear relative, or
the alteration will be found to have been gradual and
imperceptible, consisting in an exaltation or an increase

of peculiarities which were always natural or habitual.
There is also another tolerably extensive class of cases in
which the change has been subsequent to some shock
which the bodily constitution has undergone ; and this
has been either a disorder affecting the head, an attack
of paralysis, a fit of epilepsy, or some fever or inflammatory disorder.   The change, however brought
about, is always found in the temper, disposition,
habits, and moral qualities of the individual, and
is uncomplicated with any delusion or other
evidence of derangement of the intellectual faculties.   It is properly described by Hoffbauer as being *a
state in which the reason has lost its empire over the
passions, and the actions by which they are manifested,
to such a degree, that the individual can neither repress
the former nor abstain from the latter.*   It does not
follow that he may not be in the possession of his senses,
and even his usual intelligence, since, in order to resist
the impulses of the passions, it is not sufficient that the
reason should impart its counsels ; *he must have the
necessary power to obey them.*   The inebriate may judge
correctly of his actions without being in a condition to
repress his passions and to abstain from the acts of violence to which they impel him.

Dipsomania exercises a sway, *perfectly tyrannical,*
over the entire man and his actions ; every moral power
or faculty *is liable to be perverted or deranged in its
manifestations,* but those which are most prominent,
and the most frequently exhibited in the affairs and conduct of life, are the most liable to deranged actions.

"*The principle of forming volitions, and of carrying
them out into acts, must be fully possessed to render
a being accountable.*"   When, therefore, the first is
necessarily rendered incomplete, or the last prevented
by some insurmountable obstacle, all accountability is
destroyed.   It is in the first only that we witness the
agency of psycho-sensory insanity.

Inebriety is a disturbing element thrown into the very

sources whence volitions are derived, and either contributes in a large measure to the formation of those that would otherwise remain unformed, or prevents the formation of those that would otherwise be formed. In either way, it disturbs the ordinary normal operations of the mind, and thus absolves it from accountability.

It is not easy to define in what respect this new element modifies the volition or the act. The inquiry in relation to the former is unnecessary, except so far as it qualifies the latter. In regard to the manner or respect in which it modifies or affects the latter, so as to absolve from its consequences, there can never be expected an entire agreement between writers or thinkers, or even the decisions of judicial tribunals. I have supposed we might find, in *irresistibility*, a principle upon which all might agree. That wherever this quality should be found attached to an act, so far as to control it, the actor, in respect to such act, should be deemed irresponsible.

Without moral liberty there can be no responsibility for crime. *In the normal, sane state of the faculties, this enters as an essential element. In the deranged state of the moral faculties, where the sources of impulse, motive and feeling are perverted and deranged, this liberty is destoyed, and with it the accountability for actions. Irresistibility*, where it arises from deranged or perverted actions, should absolve from all accountability, because :

1st. *The act is unavoidable, and the action, therefore, is no more a subject for punishment than a machine for going wrong when one part of its machinery is out of order. To administer punishment, under such circumstances, would shock all the moral sympathies of men.*

2nd. One of the purposes of punishment would never be answered by it, viz., the reformation of the criminal. If the act be irresistible, the whole effect of punishment upon the individual must be lost.

3rd. Another of the purposes of punishment would remain equally unanswered, viz., the salutary effect to be produced by it upon the minds of others. That effect, instead of being salutary, would be in a high degree injurious, as it would shock all moral sensibilities, and create a horror of the law itself, which could thus needlessly sacrifice life without answering any good end or purpose. From this view it should follow —

23. That in any given case where dipsomania is alleged as a defense for homicide, if, at the time of the killing, the reason and mental powers of the individual who committed the deed were so deficient that he had no will, no conscience or controlling mental power; or if, at the time of the homicide, through the overwhelming violence of mental disease, his intellectual power was for the time obliterated, then in either of such cases the individual should not be held guilty of murder; or if, from the evidence in the case, it shall appear that the individual who committed the homicide was for a long time before, and at the time of the killing, laboring under mental disease, attended with delusions, and that in a paroxysm or outbreak of this disease of the mind, his reason and judgment were for a time overwhelmed and suspended, and while they were thus overwhelmed and suspended, the individual committed a homicidal act, he should not be held guilty of murder; or if by reason of mental disease an individual becomes subject to great, causeless, and violent paroxysms of rage, so that in any given case his power of distinguishing whether he is committing a crime is lost or suspended, he would not be guilty of murder if he committed a homicide, or, if in any given case the evidence shall show that from any predisposing cause an individual's mind is impaired, so that there is a prolonged change in his character, becoming sad, gloomy, and unsociable, and without interest in things he was formerly interested in, and under the influence of said causes, he becomes incapable of governing himself in reference to any particular person, and at the time of committing a homicide, is, by reason of said causes, unconscious that he is committing a crime as

to such particular person, he should not be held guilty
of murder in case of a homicidal act committed on such
person.

24. The instructions of Chief Justice Perley, of the
Supreme Court of New Hampshire, to the jury in the
case of the State v. Pike, tried October, 1876, show that
he was well versed in the principles of psychological
medicine, and believed that legal tests of responsibility
should always be derived from medical authorities who
found their opinions on the observation of facts, and
that the courts should not trespass on the province of
the expert any more as regards insanity than in the case
of consumption, yellow fever, or the contagious diseases. No one but a practical alienist and neurologist
can have a thorough and accurate knowledge of the
manifestations of mind while under the influence of disease. Such knowledge is necessarily confined to persons
who have made insanity, inebriety, and nervous diseases their special study. Are men, however eminent,
qualified to lay down general principles touching the
measure of responsibility which is left after mental disease has commenced, who never observed a single case
of inebriety closely, who know nothing of its various
forms, nor of the laws which govern its origin and
progress, and who are not practically acquainted with
the operations of the inebriate mind? In the trial
referred to, Chief Justice Perley used this language:
"That, if the killing was the offspring or product of
mental disease in the defendant, the verdict should be
'not guilty by reason of insanity;' that neither delusion nor knowledge of right and wrong, nor design or
cunning in planning and executing the killing, and
escaping or avoiding detection, nor ability to recognize
acquaintances, or to labor, or transact business, or manage affairs, is, as a matter of law, a test of disease; but
that all symptoms and all tests of mental disease are
purely matters of fact to be determined by the jury."
We should have a code which should be alike in every

State of the Union, in which it should be provided that
*no act done by a person suffering from the disease of
inebriety, or any other condition of mind in which the
person is involuntarily deprived of the consciousness
of the true nature of his acts, can be punished as an
offense.*

If the evidence of inebriety is once established, the
responsibility of the party is done away, and no persons
should be considered capable of giving opinions in judicial proceedings relative to inebriety unless they have
extraordinary knowledge and skill relative to the particular disease — inebriety — and have possessed unusual
opportunities for studying the character and conduct of
the inebriate. The plea of inebriety calls for a careful
and impartial investigation, and in criminal trials, where
inebriety is alleged as a defense, only men of experience
in the management and treatment of mental diseases and
inebriety should be allowed to examine the accused.
With respect to the way in which this should be done,
we refer the readers to our address before the New York
Medico-Legal Society on a "Plea for Lunacy Reform,"
published in the September, 1883, number of the
"Medico-Legal Journal" of New York. It will be seen
that we there advocate a commission of experts who
should, as in France, examine the accused and make a
written report of his mental condition to the court at the
time of trial.   In France, where this thing is done
properly, if the commission report the accused as insane,
there is no trial, the accused being sent to an asylum.
In all such cases the prisoner should be placed in the
State hospital for the insane or inebriates, that he might
be observed by the superintendent, who, when satisfied
as to the mental condition of the accused, should duly
report in writing respecting his mental condition.

Respecting the effect which the disease of inebriety
which exerts on *Testamentary Capacity* we refer our
readers to p. 320 of Redfield's American Cases on the
law of wills : " Drunkenness, which so far obscures the

mind and memory as to render the person incapable of doing business understandingly, disqualifies one from making a valid contract, or doing any other valid act requiring judgment and discretion. But any slight degree of intoxication caused by drink, which only slightly obscures the reason, will not have this effect. A will made by a drunken man is invalid.

Digest of opinion of the court in Peck v. Cary, 27 New York Reports, 9 (1863). p. 326, Redfield's American Cases: "Drunkenness, to incapacitate one to execute his will, must produce at the time of execution a degree of excitement depriving the person of such clear mental perceptions as to be able properly to understand the transaction. Commonly, after the effect of the stimulus subsides, he will be entirely competent to execute his will. The character of the instrument will be some guide to the testator's mind in executing it. But in order to justify the conclusion of want of testamentary capacity, it must be something more than unreasonable; it must be so violent a departure from what is just and reasonable as not to admit of satisfactory explanation short of referring it to disordered intellect."

"If fixed mental disease has supervened upon intemperate habits, the man is incompetent and irresponsible for his acts." (U. S. v. Drew, 5 Mason, C. C., 28.)

"If he is so excited by present intoxication as not to be master of himself, his legal acts are void, although he may be responsible for his crimes." This last should have added to it the following clause, "provided that the act of drinking was a voluntary act and not involuntary, the man being a dipsomaniac."

In medico-legal cases, where the defense is to be alcoholic inebriety, there should be clear evidence in the following points so far as obtainable:

1. Of hereditary tendencies.

2. Of sufficient predisposing causes, such as severe accident or disease, loss of fortune, overwork, etc.

3. Of weakness of the intellectual powers, delusions, or of epileptic fits and previous inebriate or insane acts.

4. The conformation of the cranium and the state of the general health, as well as the habits of the accused, should be carefully inquired into.

5. The *procès-verbal* or depositions should be carefully studied as to the nature of the crime, its mode of perpetration, its suddeness, the existence or otherwise of provocation and possible motives, the time, place, etc., and the behavior of the criminal afterwards, when first arrested or on giving himself into custody.

Homicidal insanity of the inebriate offers the following characters, viz.:

1. The homicidal acts of insane persons have generally been preceded by other striking peculiarities of action, noted in the conduct of these individuals, often by a total change of character.

2. They have often been discovered to have either attempted suicide or to have expressed a wish for death, or to have even wished to be executed as criminals.

3. Their acts are motiveless, or in opposition to the known influences of all human motives. A man, known to be tenderly attached to them, murders his wife and children, etc., etc.

4. Their subsequent conduct is characteristic; they seldom seek escape in flight, even deliver themselves up to justice, acknowledge their crime, describe their state of mind, or remain stupefied by that horrible consciousness of the awful deed.

5. No accomplices, no assignable inducements, and if there is premeditation, it is of a peculiar kind.

THE HERMAN MURDER CASE (for the notes of which we are indebted to T. D. Crothers, M. D., Hartford, Conn.).—Charles Herman was arrested for the murder of his wife at Buffalo, N. Y., November 1, 1885. Seven weeks later, December 21st, he was tried for this crime, found guilty and executed February 12, 1886, about fourteen weeks from the time the crime was committed.

17

A study of the evidence in this case from a scientific standpoint brings out some very interesting conclusions.

The following facts in the history of the prisoner and the homicide seemed to be unquestioned : The prisoner, a Prussian by birth, could give no history of heredity. He was about forty-two years of age, and had drank for many years, seemingly governed by no other motive except his ability or inability to pay for it. He had been married for eleven years, and lived happily with his wife up to within four or five years. When after drinking he manifested an intense suspicion of her infidelity. This had grown into a settled conviction, although there was no evidence that it was true. When sober no reference was made to this suspicion. He worked at his trade as a butcher, but changed places often, probably because of his drinking. Two years before the crime he sold out his furniture, tramped to Chicago and back, and commenced to keep house again. He was a quiet, reserved man, but when drinking talked of his wife's infidelity, and threatened to kill her in the hearing of some friends. On several occasions when drinking freely he had quarreled and attempted violence to her. He was not often stupid when drinking, but was irritable and suspicious and greatly changed in conduct and manner.

In regard to the crime, it was in evidence that he had been drinking freely every day for a week before, and although not intoxicated, was under the influence of spirits. He was known to have drank beer and spirits on the Sunday on which the crime was committed, and was seen on this day with his wife as usual. On Monday, Tuesday, and Wednesday he was noticed coming and going, drinking as usual, only his wife was not with him. The absence of his wife created suspicion, and from a search she was found in bed with her throat cut. He was arrested, and acknowledged killing his wife on Sunday evening, and placing her body in the bed and sleeping with it for three nights from Sunday to Wednesday, going away in the morning and coming back every night. He seemed to have no conception of the crime, and made no effort to escape. When arrested in a saloon he talked freely, describing the incidents of the murder, giving no reason for it, except that his wife was going out and would not stay in when he asked her, hence he threw her down and cut her throat. A few hours later, in the jail, he became restless and very nervous from the withdrawal of spirits, and could not sleep. This passed off in a few days, then he denied all memory of the past, claiming to have forgotten every detail of the murder. This he continued to assert up to death, and also manifested general indifference and unconcern about himself.

The following conclusions from these facts were fully sustained by testimony : 1st, The history of the prisoner was that of an inebriate who drank steadily whenever he could get spirits, chiefly beer and whisky. He was unthrifty, and changeable in his character and habits. His suspicions of his wife's infidelity grew with the increased use of spirits, and finally culminated in the murder. 2nd, The circumstances of the crime, the act itself, and his obliviousness to the consequences following from it ; also, his conduct at the time and later, with the absence of all reason or emotion, suggested some form of insanity. 3rd, His conduct in jail, after the first few days, when suffering from the removal of alcohol, was not unusual. He was very reticent, but acted with reasonable sanity ; the only fixed idea concerning himself was that the man he alledged to be intimate with his wife was responsible and should be in his place.

The *defense* was insanity and irresponsibility due to alcohol, and probably alcoholic trance. This was based on the history of excessive use of spirits, with the usual characteristic delusions of marital infidelity. The trance state was indicated by his conduct after the crime and general indifference of the act and its consequences ; also the automatic character of the crime, done in the same way he had been accustomed to kill animals. His first recital of the details of the crime, then loss of memory of all these events, was also characteristic of this state. His crime was probably committed in a trance state, in which he was oblivious of what he was doing, and most naturally acted automatically from an insane impulse and in a state of partial dementia.

The *prosecution* denied all evidence of insanity, and claimed that premeditation and brutality marked all the symptoms. The medical witnesses for the prosecution doubted the existence of alcoholic insanity and alcoholic trance. Two medical men were confident that spirits could be used for years to excess without causing any degree of insanity or mental impairment. One physician swore that he did not think it was the alcohol that intoxicated. The usual hypothetical questions were answered in the usual dogmatic and confused way. The possibility of insanity was doubted, because the prisoner did not then appear like an insane man.

The judge's charge to the jury entered minutely into the question of premeditation and knowledge of right and wrong and responsibility of inebriates. The letter of the law was followed closely, and the jury was told to discriminate on questions of fact and science which were clearly beyond the mental range of the judge or even the most scientific experts to determine.

The verdict was, guilty, and the prisoner manifested the same indifference to his condition up to his execution.

A review of the facts brought out on the trial appears to fully sustain the following:

1st. Charles Herman belonged to a not uncommon class who both drink and act in a way and manner that proves a defective brain and faulty judgment, also a degree of mental incompetency that should never be mistaken.

2nd. Such men always have a defective heredity and a history of neglect, bad living, bad surroundings, bad nutrition. If to this be added inebriety, mental unsoundness is always present.

3rd. The prisoner had used spirits for many years, to excess, and although not often intoxicated, he would be under the influence and exhibit the effects of spirits nearly all the time. This, of necessity, would impair his sanity and render him more or less incapable of realizing the nature and consequences of his acts. Any man of average health who uses spirits continuously for years, will have a defective brain power and brain control. Such men are practically suicidal dements, living along the border lines of pronounced insanity, and likely, any moment, to explode into wild mania. The paralyzing effect of alcohol, even in small doses, always breaks up normal brain control and conception. The terms, voluntary intoxication, and free will to abstain at pleasure, are metaphysical delusions, contradicted by all scientific study of these cases.

4th. For the past four years Herman exhibited delusions of his wife's infidelity. This, all writers agree, is a characteristic symptom of alcoholic insanity, and exists with or without any reason or basis. These delusions had grown steadily with the increased degeneration from spirits. He had most naturally talked about it to his friends, threatening vengeance, and in his inebriate state saying many things which on the trial were considered evidence of long premeditation. He had frequent quarrels with his wife and had attempted violence on several occasions, but the fact that he was not stupidly intoxicated was regarded as proof of sanity, and such acts were called willful and signs of bad temper. In reality such persons are more insane, irresponsible, and dangerous than if stupid from spirits. The fact that when sober he did not complain of his wife, and was a good-natured, quiet man, and only irritable and suspicious of her when drinking, was strong evidence of his mental aberration.

5th. The fact of drinking to excess, with peculiar delusions of his wife's infidelity, was evidence of impaired and disordered brain. No apparent sane realization of his acts or conversation could alter this fact. Such cases not unfrequently exhibit a degree of mental soundness and premeditation in thought and act, which, from a careful study, are found to be only a mask. They are the really dangerous classes of the insane, because their mental condition is more or less concealed.

6th. Herman was using spirits to excess for a week and more before the murder, and on the day of the murder he was in a mental state fully prepared for some insane act. A quarrel with his wife was most natural, resulting in a murder committed in the way he had butchered animals before. Had he manifested a realization of this act, in trying to escape or conceal the crime, or given himself up with reasons and explanations, some sanity might have been inferred. On the contrary, he seemed profoundly oblivious to what he had done, and not only threw the body on the bed, but laid down and slept three nights with the corpse. He drank during the daytime, and seemed in no way disturbed or different from what he had been before. His conduct was unmistakably that of an insane man, and one incapable of realizing the nature and character of his acts.

7th. His conduct after the crime, in giving all the details of the act, and then, when he recovered, losing all memory or recollection of them, whether real or feigned, was not that of a sane man, or one who had any clear comprehension of the nature and consequences of his acts. Alcoholic trance was a condition likely to be present. The character and manner of the crime would point to it, but his statements after, unless confirmed by other facts, would not be positive, but would only sustain a supposition of this state.

8th. His conduct and appearance in jail after the first few days might not give any indications of his real state. The absence of delusions, hallucinations, or any gross physical symptoms of insanity was by no means evidence of a sound mind or responsibility.

9th. The medical experts for the people assumed that the conduct and acts of the prisoner were sane, and that evidence of insanity must be found in his present appearance and conduct. His apparent sanity in act and thought and the general good physical condition sustained their view of full responsibility and mental health. These examinations were limited to one or two interviews of an hour or more, and were made with a view of finding some well-defined symptoms of insanity. No scientific study of the case seems to have been made. Each witness apparently brought to the case a group of symptoms by which to gauge the mental health of the prisoner. All previous conduct, unless marked by great mental aberration, was not considered.

10th. In reality, from a scientific study, if the facts of his inebriety were true, his brain was impaired. He was practically a lunatic, incompetent to judge of his acts, and his power of control and responsibility was certainly impaired. The crime and his conduct after sustained this inference. His general indifference of manner and interest in himself and future was further evidence.

The hasty trial, speedy execution, and failure to comprehend the criminal and the crime, and the pressure of public sentiment, were all inimical to justice. The progress of humanity and the cause of truth gains nothing by taking the life of a poor alcoholic imbecile, while the intelligence of the age is outraged by the application of mediæval theories of human responsibility and divine vengeance. Whatever the law may be concerning crime committed under the influence of spirits, science demands that its application shall be along the lines of natural law and observed fact. If Herman was of sound brain and capacity to realize the nature and consequences of his act, the punishment by the law was just; but if Herman possessed a defective brain and impaired consciousness of his acts, such punishment was a crime, as much so as the murder itself. Security to life and prevention of similar crime can never be secured by the injustice of taking the lives of irresponsible persons.

The hanging of insane and idiot criminals never checks

the crime of such persons. Other Hermans and Ottos* will go on committing crime just the same. The time has come for a change. The progress of science demands it. The confusion of courts, the uncertainty of juries, and the difficulty of physicians in deciding on the brain health of prisoners in a few interviews make it impossible to secure the ends of justice in such cases. Herman's case should have been the subject of scientific study before it was brought into court. The inebriety of the prisoner and the peculiar character of the crime called for special study that could not be limited to a few observations by physicians, and also could not be decided under the pressure of public feeling. The conclusion is inevitable that the supposed justice of Herman's trial and execution was grave injustice, and was only another example of judicial murder, which, unfortunately, is not uncommon in this country. The trial and execution of Herman is beyond recall, but the failure to realize the true condition of the man and the crime may serve as a landmark and warning in future cases.

---

* At the time of the trial a poor lunatic, Peter Otto, was under sentence of death for wife murder, and, although bravely defended, was finally hung. The same strange medical testimony and misconception of the case prevailed. The strange delusion that both Herman and Otto were sane pervaded the court and community, and was a strong influence in both trials. Later reflection strengthens the conviction that both cases were sad judicial blunders, in which two irresponsible men were punished as sane and responsible.

A CLASSIFICATION OF MENTAL DISEASES AS A BASIS FOR
MEDICO-LEGAL STATISTICS REGARDING THE INSANE.

*Emotional Disorders. — Intellectual Disorders (includ-
ing Disorders of the Will).*

MENTAL DEPRESSION.	MENTAL EXALTATION.	MENTAL WEAKNESS.
1. Hypochondriacal Insanity.	1. Mania. acute.	1. Idiocy, Imbecility and Cretinism (with epilepsy—without epilepsy).
2. Melancholia. acute. chronic. recurrent. puerperal. alcoholic. senile.	chronic. recurrent. alcoholic. dipsomania. senile. puerperal.	2. Chronic Mania. 3. Dementia. primary. terminal. senile.
3. Constitutional Affective Insanity (reasoning mania).	2. Monomania (chronic delirium). a. with delusions. b. with imperative conceptions.	organic { tumors. hemorrhage. softening.
4. Psycho-sensory Insanity.		4. DementiaParalytica (general paralysis of the insane).

We also hold that the student and physician should
have their attention directed particularly to the bodily
origin of the disorder, with a view to the removal of the
cause or morbid condition upon which it depends, and we
therefore present several different orders of insanity, not
claiming for each order distinctive psychological linea-
ments.

1. Epileptic Insanity.
2. Insanity of Masturbation.
3. Insanity of Pubescence.
4. Hysterical Mania.
5. Post Connubial Mania.
6. Puerperal Mania.
7. Mania of Pregnancy.
8.   "   " Lactation.
9. Climacteric Mania.
10. Ovario and Utero Mania.
11. Senile Mania.
12. Phthisical Mania.
13. Metastatic Mania.
14. Traumatic Mania.
15. Syphilitic Mania.
16. Delirium Tremens.
17. Dipsomania.
18. Mania of Alcoholism.
19. Post Febrile Mania.
20. Mania of Oxaluria and Phosphaturia.
21. General Paralysis, with Insanity.
22. Epidemic Insanity.

Homicidal or suicidal impulses may accompany any
of these forms of mania, also delusions and hallucinations.

The principle which I have attempted to carry into

effect in my classification of insanity is that a classification of mental diseases should be made to depend upon a combination of the psychical characters and of the pathological conditions.   In order to arrive at a practical diagnosis for the purpose of treatment, we must consider the psychical with the physical symptoms and the ætiology. The real problem for solution is to determine not merely the character of the mental aberration, but, as far as possible, the nature of the lesion of the brain and the nerves.

CHAPTER XII.

SPINAL CONCUSSION AND INJURY : ITS MEDICO-LEGAL RELATIONS AND SIGNIFICANCE.

IN this chapter I propose to treat of a class of injuries to the cerebro-spinal nervous system, where the effects are, as a rule, remote rather than immediate, and which, in many instances, affect the *mind* quite as seriously as the body.

The immediate constitutional state after an injury to the spine may be one of prostration and shock to the nervous system, in which we find our patient perhaps partly unconscious, with a feeble pulse and imperfect respiration. There may be complete syncope, with the pulse and respiration not perceptible, or the nervous instead of the vascular system may be principally affected, and the patient may be incoherent or perhaps comatose. Nausea, vomiting, suppression of urine, and convulsions, especially in children, may occur. The local effects in injuries of the back are not included in the scope of this chapter, and comprise wounds, contusions of every kind, fractures and dislocations. Injuries to the backs of children may occur from blows or falls, without any obvious lesion of a mechanical nature occuring, and after some time elapsing, we shall see our little patient complaining of a good deal of pain, and upon examination we shall discover redness and swelling in the tissues, and with the history of a fall or blow on the back we can at once diagnosticate the existence of caries. Finally, in these cases, an abcess forms, and after our incision, to let out the pus, we shall easily feel the carious bone with our probe. The abcess leaves fistulous openings, from which there is more or less sanious pus discharged, and the fistulous opening is surrounded by unhealthy granulations.

These cases occur from injuries to the back much more frequently, I think, than is imagined by the profession. In a violent injury to the back of an adult, the shock is the first thing that occurs, and our patient may die immediately, with no apparent mechanical lesion. If our patient survives the shock, inflammation sets in, and our patient may develop a meningitis. The symptoms of severe concussion are pallor of the face, the respiration gasping and afterwards becoming nearly normal, pupils dilated, difficulty in swallowing, pulse feeble and perhaps slow, and the consciousness muddled. The patient might appear to be feigning. We may possibly have unconsciousness. In these most severe cases the cord is contused, and if out patient dies and we make an autopsy, we may find a little red point of extravasation.

In these cases we should use for stimulants carbonate of ammonia or alcohol moderately, with bags of hot sand to the body, with friction. Hypodermics of 1-100 grain of sulphate of atropia are also indicated as the most energetic heart and respiratory stimulant. When our patient recovers from the immediate shock, he will have confusion of ideas. He may have nausea and vomiting, which are good signs, as, when the injury is very severe, our patient will not vomit, and he *may* not if the injury is very slight. If the brain is implicated, as it may be, and there is no compression, we shall find dilated pupils, a comatose condition, the respiration stertorous, and the cheeks and the alae of the nose moving at each movement of respiration. The velum palati is paralyzed, and there is rattling of mucus in the throat. The pulse is slow and laborious. If, at the end of ten days after a severe injury to the back, the patient has slight rigors and becomes partially or wholly unconscious, we are to infer that inflammation has been set up and that suppuration has taken place.   These cases are almost always fatal. We may have a paraplegia, or we may have epileptiform convulsions.   The paralysis, if it occurs, is caused by hæmorrhage of some of the vessels of the cord, coming on as reaction comes on, or, in a few days or weeks, as the result of contraction of lymph or pus.   There may possibly be a fracture without apparent mechanical lesion, and in these cases we shall get either antero-posterior or lateral deformity of the injured part of the spine.   If the injury occurs above the fourth cervical vertebra, the pressure on the phrenic nerves will give rise to death from apnœa.   In these cases the respiration is stopped either at once or in a few days.

If an injury occur to the back in the dorsal region, we find paralysis of the upper extremities and paralysis of the bowels.   We may diagnose a possible fracture, even if not apparent.   If, in an injury to the back in the lumbar region, we get paralysis of all the parts below the seat of injury, we may also diagnose a possible fracture, although not necessarily.

We should never, in suspected fractures of the spine, endeavor to get crepitus.   A fracture here will probably kill our patient, while patients not very unfrequently recover from fracture in the dorsal region.   In a suspected fracture we should put the patient on his back and make extension and counter-extension, and union may take place.   We should put our patient with any serious injury to the back on a water-bed, so as to get equable pressure, and use a catheter three or four times in the course of twenty-four hours.   If there is tendency to paralysis, a pill of croton oil, strychnia, and colocynth may be used.

Injuries to the back may be complicated with fractures of the hip, where there may be fracture of the cervix within the capsule — an extra-capsular fracture — or fracture of the brim of the acetabulum or of the floor of the acetabulum. We may possibly get, as a complication, dislocation of the hip. There is no class of surgical injuries of more interest to the neurologist and to the general practitioner who is interested in diseases of the nervous system than those which come under the head of injuries to the spine and spinal concussion.

No injury to the spine, however slight, arising from shocks to the body generally, as in railway accidents, or from the ordinary accidents we meet with in general practice, comprising falls, blows, being thrown from a carriage, is too trivial to be overlooked, as the spinal cord may be functionally disturbed, and even organically diseased from any and all such shocks and injuries. We may have local and constitutional, immediate and remote effects from these injuries. The primary effects of a concussion of the spinal cord are due to molecular changes in the structure of the cord, while the secondary effects are of an inflammatory character, consisting of meningo-myelitis, disturbances of nutrition, with great mental and moral depression.

There is often change of character, irritability of temper, and often impairment of some of the special senses. Death may occur after chronic inflammation of the membranes and cord, lasting for three or four years, during which time our patient's health has gradually been breaking down, with slow extension of paralytic symptoms. The symptoms may be immediate or they may develop slowly after an interval of some months. In a direct injury to the spine we may find our patient with a bruise on his back, with pain on pressure, with consciousness intact, with partial paraplegia, and with more or less numbness. Febrile reaction sets in and lasts for a few days, during which time he may not be able to empty his bladder, necessitating the use of the catheter every six hours. We may find a great latitude as to the extent, degree and relative amount of paralysis of motion and of sensation in any given case. If the direct blow is on the dorsal or lumbar vertebræ, paraplegia usually results.

Sensation is necessarily affected, spasm and rigidity of the muscles may occur, the sphincters may be involved, and we shall have much pain. There may be incontinence of urine or there may be partial or complete retention. Low temperature is the rule in spinal injuries, a high temperature when we meet with it being indicative of inflammatory troubles. When we find a fatal result, it is due to hæmorrhage, laceration, extravasation, or to inflammatory softening, and our patient's recovery may be complete or incomplete. A considerable length of time may elapse between a spinal concussion or injury to the back and the development of the symptoms of the injury, which may be so slight, perhaps, as to attract little attention at the time ; just as we have seen, in a previous part of this work, that brain tissue degenerations and mental diseases may be separated by long intervals of time from the too premature and intense stimulation of the brain, which causes these nerve and brain diseases. This is a very important medico-legal point. The muscular, tendinous and ligamentous structures of the spinal column may be very violently wrenched and sprained by injury or concussion, without injury to the cord itself. These cases may recover in a few weeks, or, in delicate persons, they may lay the foundation for serious organic disease. If inflammation is developed in the fibrous structures, it may extend to the meninges of the cord, and this possible danger should not be overlooked or ignored in our prognosis. Our patient may slowly develop cerebral symptoms from the extension upwards of meningeal irritation. After a spinal concussion it is not at all uncommon for our patient to undergo a gradual change, both mentally and physically, and he is never the same man again. He gradually becomes an invalid, unable to apply his mind to business, or to stand the ordinary cares and worries of life, which previous to his injury had never troubled him. There is

decided mental failure, which may proceed to complete
imbecility or insanity.   The mental responsibility of
such a person is greatly lessened, and his testamentary
capacity may be also affected.   When injury of the back
is severe enough to produce, at the time of injury, unconsciousness, insensibility, stupor or syncope, then the
severity of the concussion is such as to produce an immediate injury of the gravest nature to the central nervous
system, and *never* afterwards does such an individual
have complete restoration to health.   After a concussion
of the spine, many weeks or months may elapse before
the more positive and distressing symptoms occur.   In
the interval, however, our patient suffers from poor
health, his nervous power has gone, and his face is
anxious and careworn.   His memory is defective, his
thoughts are confused, his business aptitude is lost, his
temper is changed, his sleep is poor, and his special
senses impaired.   There is also loss of motor power and
a diminution of sensation in the limbs.   The patient at
first complains of weariness on slight exertion, either
mental or physical, followed by the modifications in
sensibility, pain and rigidity of spine, cerebral disturbance, and, as I have remarked, loss of motor power.
When there is myelitis, the sensibility is at first aug-
mented, and then, as the myelitis becomes chronic, the
gait is very much affected.   Whether acute or chronic,
myelitis is much more apt to attack the lower portions
of the cord, and we get cerebral complications, we may
be sure we have more or less spinal meningitis.   The
coexistence of meningitis and myelitis is what we generally find in our patients who have suffered from severe
injuries to the spine.   I think it is rare to find inflamma-
tions of the spinal membranes limited to the spinal canal,
and that there is an extension of the morbid process
which gives us, as a result, an increased vascularity and
inflammation of the arachnoid.   In spinal meningitis we
have, as the most marked symptom, severe pain along
the spine and down both legs.   These attacks of pain

may be separated by intervals of almost complete ease and comfort. The pain is soon accompanied by stiffness of the muscles of the back and legs. Any movement of the body, neck or legs, gives rise to pain. There is absence of paralysis, some exaltation of sensibility, loss of power over the bladder, and partial loss of power over the bowels. There is absence of spinal tenderness, and there is also an absence of marked spasmodic symptoms. In proportion as the higher portions of the cord are affected, there is difficulty of mastication and deglutition. Difficulty of breathing generally is present. There is a little sympathetic fever, and there may or may not be cerebral symptoms. There is no increased reflex excitability.

Myelitis is characterized by paraplegic anæsthesia ushered in by tingling in the parts, which soon become anæsthetic. The paraplegia is preceded by restlessness, rather than by more marked symptoms. There is a very uncomfortable feeling of tightness around the waist or elsewhere, as a constant symptom of myelitis. There is, as a rule, absence of pain, except when our patient is suffering from the combined meningo-myelitis, of which I shall presently speak. In simple myelitis, I do not think there is much pain in the spine or extremities. There is an absence of spasmodic symptoms. As a very early symptom there is a want of control over the bladder, which depends upon a paralysis of the accelerator urinal and compressor urethral muscles. There is a want of control over the rectum also, caused by paralysis of the sphincter ani. There is absence of tenderness on pressure in any part of the spine. There is an altered sensibility to heat and cold, by which a feeling of burning is felt when a sponge soaked in moderately warm water, or a piece of ice is applied immediately above the seat of inflammation. There is annihilation of reflex excitability. There is a diminution of electro-motility and electro-sensibility in the paralyzed muscles. There may or may not be priapism. The urine is generally

18

alkaline, but neither always or necessarily so.   There is
marked difficulty in breathing.   The state of the circulation is asthenic.   There is a tendency to bed sores, and
there is in simple myelitis absence of head symptoms.
In a patient who has suffered a severe injury to his back,
we very probably may have coexisting cerebral meningitis, spinal meningitis and myelitis, and the symptoms
will be those of meningitis or myelitis, as the one or the
other preponderates.   Our patient, finally, as the result
of nervous shock from an injury to his back, may escape
organic trouble but develop spinal anæmia and marked
hysteria, lasting many months.   Meningo-myelitis is a
very grave disease, and one which devitalizes the whole
system.   If our patient recovers, he is probably a broken
down man, and we must hereafter keep him on cod-liver
oil, phosphorus, arsenic and bichloride of mercury, with
electricity to improve his general nutrition.   A patient
who has had a spinal injury may have his vision very
materially impaired.   There may be a weakness of sight,
an intolerance of light, double vision, amblyopia, paralysis of accommodation, and anomalies of refraction.

These optic lesions are due to extension of meningeal
trouble to the cerebrum.   If the brain is unaffected, the
impairment of sight may be due to the action of the
sympathetic nerves.   The filaments of the sympathetic
that supply the eye take their origin from that part of
the spinal cord which is contiguous to the origin of the
first pair of dorsal nerves, and the portion of the cord
which extends from the fifth cervical to the sixth dorsal
vertebra possesses a distinct influence on the eyes and
vision.   I consider it certain, therefore, that we get an
affection of the optic disc and its vicinity from the various disturbances of the spine consequent upon injuries
to the back.   These optic lesions are principally due
to a cerebral meningitis that commenced as a spinal
meningitis.   We have perverted, impaired, or lost sensibility of the optic nervous tract as the result of spinal
concussion, with atrophy of the optic disc as the final

stage.   Where the brain is unaffected, the loss of sight
is due therefore to the transmission of the morbid action
from the cord to the vessels of the eye by the agency of
the sympathetic nerve, rather than by extension of
inflammation.

The medico-legal aspect of a case where there has been
a severe injury to the back, causing concussion of the
spine, should be stated by the physician who is applied
to for information, very decidedly, but briefly.   Our
patient's mental and physical vigor are gone, and, if the
changes have been organic in the cord and brain, gone
probably forever.   He never can be the same man as
before the injury.   Death is far preferable to a life of
hopeless invalidism, as many such patients must ever
after lead.   The prognosis in these cases is always very
grave if, after a year or two has elapsed from the time of
the occurrence of the accident, the symptoms of meningo-
myelitis either continue to be gradually progressive, or,
after an interval of quiescence, suddenly assume an
increased activity.   Cases of injuries to the back, without apparent mechanical lesion, may die; *first*, at an
early period, by the severity of the direct injury; and
second, at a more remote date, by the occurrence of
inflammation of the cord and its membranes; and finally,
after the lapse of several years, by the slow and progressive development of structural changes in the cord and
its membranes.   The patient, if he does not die, may have
a mitigation of his symptoms—an amelioration—but a
thorough cure, after severe spinal concussion, we shall
never, or very rarely, witness.

Our prognosis in these cases should always be very
guarded.   The chances are decidedly against our patient
as regards complete recovery.   The general health tends
progressively to break down, and if our patient gets up a
chronic myelitis, the chances are that he will die in a few
years.   Those cases are the most favorable in which the
symptoms attain their intensity *soon after the injury*,
while a long interval between the receipt of the injury

and the development of the spinal symptoms is unfavorable to our patient's recovery. The treatment is rest, counter-irritation, nerve sedatives, and the constant current of electricity to the spine. For a constitutional treatment I prefer iodide of potassium, quinine, and bichloride of mercury, with cod-liver oil. The constant current is indicated when our patient has developed a spinal anæmia, and the phosphide of zinc and strychnia are valuable also. In inflammatory states of the cord electricity and strychnia would be contraindicated, while ice-bags and ergot would do good. We must give our patient cheerful surroundings, and build him up physically and mentally, and in exceptional instances we may see complete recovery.

In conclusion : 1. It is important, from a medico-legal point of view, to remember that from an injury to the back we may have unsuspected fractures of some of the vertebræ ; and that, although there may be no head symptoms and no head injury, and no paralysis, yet the injury inflicted may be of a fatal nature, although life may be prolonged for several days until death occurs from some accidental movement.

2. We may also have injuries to the back or spine occurring that are necessarily fatal, without any direct blow on the spine, but from falls on the head. We may get an inflammatory softening and disintegration of the cord in such cases.

3. We may have many diverse kinds or varieties in the extent and degree of paralysis of motion and sensation. Of course the symptoms in any given case will be varied in character and extent, according to the location of the injury, the force with which it has been inflicted, and the amount of organic lesion that the delicate substance of the spinal cord has suffered from by the shock or jar that has been inflicted upon it.

4. We may have a severe contusion, with paraplegia and an unsuspected laceration of the intervertebral ligaments, followed by death in a few days.

5. We may have a slowly developed spinal meningitis from a direct injury to the back in railway collisions, terminating eventually in death.

6. We may have a direct injury to the back and slow developement of paralytic symptoms.

7. We may have compression and concussion of the cervical spine from a blow on the head, with paraplegia and a slow recovery.

8. We may have falls from horseback, or from carriage accidents, with concussion of the spine, immediate paralysis and complete recovery.

9. We may have a direct injury to the back, without apparent mechanical lesion, followed by a paralysis of one limb only.

10. We may have a concussion of the spine from falls on the back, followed by partial paralysis of sensation and motion of the lower limbs, without affection of the sphincters, and terminating in incomplete recovery.

11. We may have falls on the back, with partial paraplegia and recovery.

12. We may have cases of slight injury to the head or back, followed by serious, persistent, or fatal results.

13. We may have epilepsy, appearing by transmission in children, whose parents have become epileptic by an injury to the spinal cord.

14. We may have hyperæsthesia, anæsthesia, pain, and perverted sensations of all sorts and kinds in cases of spinal concussion from injuries to the back.

15. After an injury to the back we may have complete recovery, incomplete recovery, permanent disease of the spinal cord, and meningitis, or, finally, death.

16. We may have a terrible nervous shock resulting from injuries to the back, no immediate effects, a chronic meningitis of the cord and base of the brain, and an imperfect recovery.

17. We may have a violent fall, with no injury externally apparent on the back or head, in which the patient

is much shaken, develops symptoms of concussion of the spine, and makes a very slow recovery.

18. These general shocks, with symptoms of spinal concussion and meningitis, are generally the result of a railway collision.

19. We may have sprains or violent wrenches of the back or spine, followed by every variety of harm to the spinal column, ligaments, the cord, or its membranes.

During the past summer I had under treatment a case of meningo-myelitis, the result of a blow on the spine received, so far as we could tell, some months previous. The patient was a lady of 27 years of age, who for some years had been addicted to periodical indulgence in stimulants, which had got her tissues into the worst possible state to resist any injury. When I was first called to see her, her pupils were somewhat dilated, the gait staggering there was confusion of mind, great irritability, and not a perfect understanding of her condition and surroundings. These symptoms increased, the bladder became affected so that the catheter had to be used thrice daily, the memory was very defective, there was restless sleep with nocturnal delirium, vision was impaired, and the patient saw sparks and flashes of light, the head was hot, and she would awake out of sleep greatly frightened. Sensation was absent in both legs, and, as the case progressed, all reflex movements disappeared. There was soon complete paraplegia, with numbness of a distressing nature. There was pain on pressure and on movement over the lumbar vertebra. Alteratives, and tonics, and rest, with counter-irritation, were all unavailing. The examination of the spine by the hot sponge showed exalted sensibility and pain at the level of the inflammation. In walking, the patient, in the first stages of the disease, kept her feet apart and straddled, and she had a distressing sensation as if a cord were tied tightly around the waist, and complained of the same sensation in the limbs. She also complained of shooting pains in the limbs, and of great coldness of the feet. She lost weight rapidly, even when eating well and taking no exercise. The pulse at first was slow and never rose above 120, and the temperature remained nearly normal until it finally rose to 160°, with an irregular and intermitting pulse of 120. Upon close inquiry I could not learn that in this case there had, since the accident and the time (about four or six months) of the supervention of the serious symptoms, been any interval, however short, of complete health. The cerebral disturbance in this case, the headache, confusion of thought, loss of memory, and defective vision were referable, I think, to cerebral meningitis and arachnitis. This appeared to be the result of the inflammation of the cord and its membranes ascending so as to involve the intracranial organs.

Bampfield, in his prize essay on the spine, page 175, says: "Concussions of the spinal marrow sometimes occasion extravasation of blood, or bloody serum, within or without the theca vertebralis; and, in addition to the usual loss of sensation and volition in the extremities, the respiration shall be disordered. These cases are frequently fatal. Dissection has disclosed extravasation of blood, in one or more clots or patches, between the spinal cord and the spinal dura mater, or in the cellular texture investing that membrane. All the membranes of the spinal cord have been found either separately torn up or ruptured, or the whole have been lacerated, and have allowed the medulla spinalis to protrude like a hernia, or to escape through the rent in a mashed condition."

Concussion of the spinal cord is distinctly treated of in "Dr. Gross' System of Surgery," Vol. II, page 247, and in "Holmes' System of Surgery," Vol. II, page 238. It is not only unwise but unjust to sneer at all injuries to the nervous system, except such as produce, almost immediately, the coarsest and most visible lesions. Perhaps some of my readers may be unaware of the fact that the late Charles Dickens was never the

same man after the railway accident which he
described and from which at the time he seemed to
suffer but little. The fair rule to follow is that laid
down by Woodman and Tidy in their "Forensic Medicine," page 811, where they say that there should be
some physical facts, *i. e.*, some symptoms cognizable by
the senses, which can be adduced in support of the com-
plainant's statements. Besides paralyzed limbs, we may
have affections of the special senses, vaso motor paral-
yses — such as unilateral or localized sweatings and
flushings — irregular distribution of temperature, etc. ;
affection of the pulse, a quickening of 100 or more,
certain neuroses, such as choreic movements, locomotor
ataxy, hysterical and epileptiform fits, tremors, etc. ;
affection of the speech and handwriting, change of
character and conduct ; disturbance of digestion and
nutrition, showed by wasting, constipation, etc. ; sleeplessness, affections of the urine, sugar, and albumen with
blood, change of color of hair, and quite rarely
abscesses, aneurisms or tumors may be developed by the
shock of the collision. Whilst we deprecate overesti-
mating injuries and the giving of undue damages, railway companies should certainly be held liable for
railway accidents, which may very readily produce
serious injuries to the nervous system.

CHAPTER XIII.

THE PSYCHOLOGY OF CRIME.

PERHAPS there is no question which interests physicians and jurists alike more than does the question of the degree of responsibility which attaches to the class of the mentally unsound who, "laboring under the tyranny of a bad organization," are constantly or periodically impelled to commit crime. This, as we shall prove further by the admirable researches of the distinguished Professor Moriz Benedikt, of Germany, is the result of a pathological state of the brain, connected with a peculiar type of skull development (shortening of the occiput, anterior vertex steepness, "*scheitel-steilheit*," and then, in decreasing progression in the asymmetry and the flattening of the occiput). Examination of the brains of murderers after death support these facts. There is a resemblance to the brute in some of these brains, in that the cerebellum is not covered by the occipital lobes and that there is also a deficient development. The insane, as a rule, never exhibit remorse or feeling of guilt for crimes committed. This is a well-known psychological truth. It is a fact not so well known, but which is supported by the assertions of Benedikt and Holtzendorff, that great criminals also rarely are penetrated by a feeling of their guilt or exhibit remorse. If they do, it is only temporary. The psychological state of criminals who exhibit the anterior vertex steepness is the analogue of recurrent mania or of chronic mania with lucid intervals. It is also analogous to states of disease in which attacks of illness of more or less short duration alternate with more or less long, and generally for a time preponderant healthy intermissions. They relapse periodically into crime, and, they all unanimously testify, from an irresistible impulse. I have a patient belonging to an old and good family who is thus ethically degenerate, in marked contrast to all the rest of the family. His history is that, when a child, he would lie about everything with apparently no reason, would not take a healthy normal interest in the sports of boyhood, but was very lazy; did not like to study, and, when older, would not work; never drank, but was addicted to self-abuse. When older, he wandered aimlessly from place to place, generally making himself so obnoxious by his foolish conduct that he shortly wore out his welcome.

Upon coming under our care we discovered at once the peculiar criminal skull formation and observed that he was, like the insane, much affected by barometrical states, and especially by thunder-storms. He seemed to be ethically degenerate, and there was a moral imbecility in his case which was probably congenital. He repeatedly has told me that he felt afraid of himself and acknowledged that he had criminal impulses, but would not state explicitly in what direction they lay. He claimed that these impulses occurred only at times, and that they were irresistible. Such cases are incurable when congenital, and if all of them could be sent to an asylum for the chronic insane for life much harm would be saved to the community, many crimes would not be committed, and punishment by death could, as Holtzendorff and Benedikt have shown, be practically done away with. Relapses are absolutely certain in these cases, because the brain is abnormally developed. If we could examine such a brain we should find an abnormal prominence, and also a preponderance of the fissures, which is a sign of arrested development, for, as **Professor** Benedikt has shown, this condition arises from the circumstance that certain convolutions remain stationary in the deep parts, and have therefore not arrived at their full development or have not developed themselves. We might also find that the cerebellum was not covered by the occipital lobes of the brain, as it normally is. Professor Benedikt found this in three brains of murderers which he examined post-mortem, and in a fourth case an equivalent condition was observed by him.

In an address of Professor Benedikt's before the meeting of the Juridicial Society of Vienna, December 28th, 1875, he exhibited some varieties of skull formation which play a great part in the natural history of crime, and spoke as follows concerning them: "If, in the normal skull, in a straight line from before backwards, the distance is measured from the fossa behind the auditory foramen to the most posterior eminence of the occipit, it will be found to amount to two-fifths and more of the straight line, drawn from before backwards, in the middle line between the forehead and the summit of the occiput (the sagittal diameter). I show you now that in other skulls this is not the case, inasmuch as the first line reaches one-third or one-fourth or less of the second. I call this 'brachycephalia occipitalis.' In the second place I show you that the difference in height between the highest point of the forehead and the crown of the head is but small (one and one-half centimetres). In many skulls the difference is considerable (as much as seven centimetres), and this proportion I call 'anterior vertex steepness.'"

Scheitelsteilheit. A further variety is the asymmetry of the two halves of the skull; and, lastly, please to observe the form of the posterior surface; it is in certain skulls very flat, while in others this occipital flatness is wanting.

The professional robber, Professor Benedikt thinks, is affected with ethical idiocy. Covetousness, ethical weakness of mind, pleasure in the imaginary or actual conviction of obtaining the desired means of existence without work, when mental or bodily power is deficient, or the dislike of taking this power any longer into account, are the factors, he says, out of which the psychological product of assassination for the love of gain is composed.

Violence of temperament, continuance of a strongly-excited dislike, overweening feeling of power, and of pleasure in exercising strength over relative weakness of intellect, and want of ethical development, form the psychological basis of rough manslaughter, he says, as well as of murder from revenge with slight motives. The psychology of theft he describes as excessive pleasure in reveling, and disgust for work, which form the peculiar basis of the ordinary thief's nature. These are the impulses, he says, which cause the consciousness of the balance between meum and tuum to be disturbed and finally to disappear altogether. The kleptomania of hysterical persons he speaks of as worthy of observation, in whom there is an impulse to possess everything without making use of it. Benedikt says that the whole psychological I is affected in the thief, but the ethical and the motor I, and the intellectual, in a more limited sense. He speaks of the special bank-note forger as belonging to that type of criminals who very generally relapse, and very truly says that the same prominent characteristic feature of motive ingenuity will protect a man from the path of crime if he has the talent of conception and the spirit of origination, or if a developed ethical talent is present in his disposition. The knowledge of the complicated nature of the psychology of crime is, however, extraordinarily important, says Benedikt, in the question of the degree of punishment to be awarded and of the possibility of amendment. When any one with a fierce temperament and an arrogant consciousness of strength has been mentally ill-developed, has learned only the roughest hand labor, and has not been educated in morals, he may become a useful member of society if his intellect and his cleverness are developed and the slumbering better feelings are awakened. Then is the individual, he

says, further developed, and the restraints which were
formerly wanting may now come into activity. When
the conditions are of this nature that from the impulses
leading to crime there is no dissuasion and to those
restraining from it there is no persuasion, there is
no chance of improvement, and legislative punishment becomes stronger and stronger for habitual criminals. There is then, Benedikt says, no advantage in
setting such a criminal free, for he will again commit
crime. If we now make an inquiry on the ground of
these empirical experiences and their analysis, in order
to find whether, in a certain percentage of certain
grades and categories of crimes, certain changes cannot
be detected in the brain or the skull, we shall find that
we do not need to seek, as the old doctrine of Gall
attempted to do, for the foundation of crime in altogether local developmental alterations, but that excesses
and defects of constitution and development must be
present in the three great centres of ideas, of motion and
of sensation. Benedikt further says that we must not
assume, because characteristic changes are present in
criminal natures, that men so constituted must necessarily commit crime. The question here, he says, is only
as to a *predisposition*, just as persons with a narrow
chest have a predisposition to tuberculosis or children of
insane parents have an insane diathesis. It must always
depend, he says, on a number of conditions whether a
nature predisposed to crime will actually become a criminal, and the clearer we are as to the psychological and
anthropological marks by which the disposition may be
revealed, the more surely shall we prevent crime by education and watchfulness. The numerical results of the
examination of a large number of heads show that
"*brachycephalia occipitalis*," while wanting in 93.5 of
normal skulls, is wanting in only 23 per cent. of robber-
murderers and 45 per cent. of murderers from motives.
It is great in only 2 per cent. of normal skulls.

Occipital flatness is wanting in 58 per cent. of normal

skulls, in only 16 per cent. of the heads of robber-murderers, and in 28 per cent. of the heads of murderers from motives.   It is well marked in only 12 per cent. of normal skulls, and in 59 per cent. of the heads of robber-murderers.   Asymmetry is wanting in 62 per cent. of normal heads, in only 10 per cent. of the heads of robber-murderers, and in 25 per cent. of the heads of murderers from motives.   It is great in only 13 per cent. of normal heads, in 43 per cent. of the heads of thieves, and in 32 per cent. of the heads of murderers.

Vertex steepness is wanting in 85.2 per cent. of normal heads, and in only 40 per cent. of the heads of thieves. Professor Benedikt says, that wherever abnormalities occur in a high degree and in combination, there exists a relapse into an earlier stage of the development of mankind, and the examinations of brains confirm this view.

The late Dr. Ray, who was an eminent alienist, said : "Let me also say that the moral pathology to be learned in these establishments (institutions for the cure of mental diseases), will have an important bearing on some of the prominent questions of moral and social science.   If we are ever to obtain a correct theory of human conduct, to discover, in any degree, the secret springs of action, or to penetrate into the mysteries of human delinquency, it must be by the study of morbid psychology in that broad and liberal manner which is possible only amid large collections of the insane.   No one who declines to receive his opinions on trust can help being embarrassed by the problems presented by many an historical name, or those revelations of character so often found on the records of our courts.   We seek in vain for any light on the questions thus raised, and are obliged to rest helplessly in the conviction that there are more things in heaven and earth than are dreamt of in our philosophies. Indeed, these difficulties cannot be overcome by any theories of human conduct which suppose the mind to be in a perfect normal condition.   They point to imperfection, or deficiency, or obliquity, the result of organic

influences, and they can be cleared up in no degree
except by the profound study of organic conditions in
connection with abnormal mental phenomena. From
this kind of study we may justly expect that a light will
be thrown on the field of history and biography, by
which many of their pages will be read with sentiments
very different from those which they now inspire. It
would show us that much of what the world calls genius
is the result of a morbid organic activity ; that many a
saint, or hero, or martyr, became such more by virtue of
a peculiar temperament than of a profound sense of
moral or religious obligation ; that the horrible crimes
which have imparted an infamous distinction to the
Tiberiuses and Caligulas of history proceeded rather from
cerebral disorder than a native thirst for blood."

Dr. Ray says elsewhere : "The researches of Gall and
Spurzheim first led to more philosophical views respecting the constitution of the brain, for although their
system has failed to obtain any considerable belief, yet
their particular proposition, that size is a measure of
power, will scarcely be disputed now. The next step, of
little less importance, was made by their followers in
explaining the apparent exceptions to the rule, by supposing a diversity of quality in the materials of which
the brain is composed. At a later period, the deterior-
ating influences of vicious or unhealthy habits and
usages were made the subject of an admirable work by
Morel, while the effect of nervous disorders on the cerebral organism was investigated by Moreau de Tours with
remarkable acuteness. The result of these and other
kindred inquiries was to establish beyond a reasonable
doubt the principle, that the brain comes into the world
with the same imperfections and deficiencies, the same
irresistible tendencies to disease or perversity of action,
which have long been observed in regard to other organs.
Thus was opened a new realm of inquiry, of unprecedented interest to the student of pathological psychology,
and of immense importance in many practical relations

of life. We have as yet but a faint idea of its full significance, but it needs no great faith to believe that it is
destined to modify very much our present theories of
human action, and throw new light on many dark problems of human conduct.   Recent investigations have
added new difficulties to a subject already regarded with
much diversity of opinion. If overt disease, manifested
by appreciable symptoms during life and various lesions
after death, can annul responsibility, the question inevitably follows, *whether that cerebral condition,—neither
of health nor of disease, as those terms are usually
understood,—which is produced by tendencies to disease or ancestral vices, may not impair it, in some
degree, under some circumstances?** This is the question of questions presented to the psychologists of our
times, and destined, undoubtedly, to raise sharper con-
flicts than any other in the whole range of medical
jurisprudence. It is involved in obscurity, it is met by
the bitter prejudices of those who lead public opinion,
and extensive investigations and various knowledge are
needed for its solution."

A grave moral impropriety, the result of criminal
impulses, is popularly called wickedness. A grave
intellectual impropriety often indicates, to even a casual observer, unequivocal insanity, and is attributed to
mental defect. Both may be equally the mental manifestations of imperfection, congenital defect, or abnormal depreciation of the cerebral system. The best proof
of this is, that insanity and crime may both appear
either in the same generation or in different generations
of the same family. This is a well known fact. "To
say," says Dr. Ray, "that a man's character and conduct are determined, in a great degree, by the original
constitution of his brain and nervous system, is to utter
a truth that can hardly be called new. Few, however,
are disposed to make any proper account of those cere-

---

* Italics are mine.

bral qualities which imply a deviation of some kind or other from the line of healthy action. It is not in accordance with the philosophy of our times to see in them an explanation of these strange and curious traits which are utterly inexplicable on the principles that govern the conduct of ordinary men. How, then, could they expect the popular approbation who find in them a clue to some of the mysteries of human delinquency? But the teachings of science, the stern facts of observation, cannot be disregarded. Whether we ignore them or not, sooner or later their full significance will be triumphantly acknowledged. In the popular apprehension, even downright insanity is regarded as of little practical account, unless it courts observation by the force and variety of its manifestations. Only its more demonstrative forms are supposed to be capable of affecting the legal responsibility of men. The world is reluctant to believe that a person, who, in most respects, is rational and observant of the ordinary proprieties of life, can be so completely under the influence of disease as to be irresponsible for any of his acts. If the world is reluctant to allow to this class of persons the immunities of insanity, it could hardly be expected to treat, with any degree of favor, those traits or conditions of mind which imply not disease, perhaps, but abnormal imperfection of the brain. And yet it cannot be denied that the course of thought, the sense of moral distinctions, the actual conduct, may be greatly affected by the influence of such imperfection. Are we not bound, then, by a sense of justice and the claims of science, to make some account of it in forming our estimates of character and fixing the limits of responsibility? Can we do otherwise without the greatest inconsistency? Knowing that an individual is descended from a line of progenitors abounding in every form of nervous disorders, shall we think it strange that some vestige thereof should have come to him? And knowing that the quality of the brain is necessarily affected by such disorder, shall

19

we not seek, in this fact, for an explanation of what
would be inexplicable upon any ordinary principles of
human conduct?"

If there is a tendency to disease, not disease itself in
any particular brain, the accumulated results of experience of many able observers all tend to show that,—1st,
it may die out; 2nd, that it may manifest itself in all
forms of nervous diseases up to fully developed mania;
or, 3rd, that it may show itself in inebriety or a proclivity to crime. The offspring of insane parents
may be insane, inebriates, or criminals. The moral
sentiments are just as apt to be affected by cerebral
defect as is the purely intellectual part of our nature.
It may be mental capacity and vigor, or it may be
moral capacity and vigor, which is attacked by disease
in any given case. Given, a *latent tendency* to disease from congenital or acquired vices of cerebral
conformation or nutrition, and no psychologist or alienist
can predicate with any certainty whether the fully formed,
fully developed attack will fall on the purely intellectual
or on the moral side of a man's nature. A great criminal
or a raving maniac may be the result of the evolution of
the morbid psychic force. How can we as far as possible
antagonize these latent tendencies, if we suspect them to
exist? I would answer, by a good physical education
and a sound mental discipline, to strengthen the powers
of the mind and keep them in healthy channels of thought,
feeling and action. In deciding between depravity and
mental infirmity, we must remember that embryonic
mental disease may leap into sudden and overpowering
activity with just as little warning as a stroke of paralysis
may result from a family tendency to it, and that an
appalling crime or an attack of furious mania may follow
close on a short initiatory period of depression of spirits.
There are cases of incurable chronic insanity, familiar to
every alienist, where there are lucid intervals simulating
recovery, where there is a resumption of apparently
perfectly healthy mental action. A casual observer

would say the person was perfectly sane. The disease, however, from what cause we know not, is merely latent, and we have personally seen the most terrific burst of mania following, with no warning save that of one sleepless night, a mental calm so deceptive that it appeared like perfect sanity. Just so in the criminal; the impulse lies in embryo, strictly in accordance with the laws of morbid action as evinced in mental disease, and we cannot tell what will be the mode of its operation. We do know that, owing to cerebral defect, it will recur quite regularly, and it seems hardly just to release a criminal who must, from the very nature of things, commit fresh crimes, and then take his life by capital punishment, when his cerebral conformation would suggest that an asylum for the criminal insane for life would be the appropriate place for him. The Townley case in England, some years ago, is an instance of an appalling crime being committed by one inheriting a tendency to insanity, where great injustice was done, and humanity and science lost sight of, because the public deprecated any judicial mercy, and penal servitude for life in Australia, with the subsequent suicide of the prisoner, completed the history of the unfortunate young man.

The facts of criminal psychology, Professor Benedikt has shown us, lead us to regard the impulse of criminal natures in the light of natural laws, and to believe that a deficient organization occasions the disposition to an abnormal moral constitution. In all four of the brains of murderers which Professor Bendikt examined, it was found that there existed a deficient anthropological development, and in all four cases sentence of death was passed, on the ground of the existence of a full responsibility recognized by judges and medical men. The existence of cerebral abnormities was in each case the cause of the criminal impulse. Unhealthy psychical function means generally, if not always, either a congenital or an acquired vice of conformation or nutrition of the cerebral system, and this fact should warn judges and juries to exercise great caution.

Professor Benedikt gives the description of the brains of two murderers who committed a murder for hire:

In one, the cerebellum is not covered by the occipital lobes, and the occipital brachycephalia is present on the left side. In the right hemisphere the ascending posterior spur is merged with the ascending part of the interparietal fissure, and reaches at the median surface into the gyrus fornicatus. The second parietal lobe is divided from the first temporal lobe by a long fissure (parietotemporal fissure), which is lost in an operculum with *plis de passage* (parieto-temporal operculum), which is bounded by the lobulus tuberis and by lobules which are probably to be regarded as processes of the first and second temporal lobes, but are pretty clearly distinguished from them. The ill-developed gyri fusiformes and lingualis rise upwards steeply from the plane of the gyrus uncinatus toward the summit of the occiput, and thereby, that the fissura calcarina may reach to the same point, is the gyrus rectus of the occiput reduced to a minimum. The vertical occipital fissure is in direct communication on the one hand with the horizontal occipital fissure, and on the other hand with the sulcus hippocampi. The horizontal occipital fissure exhibits the ape-form with its *plis de passage*. The first three temporal convolutions are arranged concentrically according to the brute type, with the concavity downwards. The gyrus uncinatus and gyrus hippocampi are very deficient in convolutions. The fissura praecentralis reaches to the middle border. (The first and second frontal fissures are divided from it by imperfect convolutions.) *Plis de passage* are found both in the horizontal occipital fissure and in the parieto-temporal operculum, in the posterior spur of the fissure of the fossa Sylvii combined with the interparietal fissure, in the anterior ascending spur of the same fissure, and also in all the furrows. On the left, the posterior ascending spur of the fissure of the fossa Sylvii rises to the median line, and the interparietal and parieto-temporal fissures are separated from it by ill-developed portions of convolutions. The vertical occipital fissure is not connected with the interparietal fissure. The second frontal convolution is crossed with the fissura præcentralis, and thereby are the first and second frontal convolutions divided into (smaller) posterior and (larger) anterior halves. The fissura præcentralis reaches as far as the inferior border, and is excessively convoluted. The first and second temporal furrows are arranged in a more normal manner. In other respects the arrangement is the same as on the right.

In the second brain, the body of the organ appeared, on the other hand, altogether oblique and shortened, and the obliquely situated cerebellum was deeply imbedded in the niche of the fourth and fifth temporal convolutions, whereby the mass of the occiput appeared to be deteriorated in a high degree behind the gyrus uncinatus. On the right, the posterior spur of the fissure of the fossa Sylvii rises high up, but without reaching the median border. The interparietal fissure is in normal proportion with it, but stands, nevertheless, in direct communication with the fissura Rolandi. The horizontal occipital fissure shows no decided similarity to the apes. The parieto-temporal fissure is not much developed; the parieto-temporal operculum is well marked. The first three temporal fissures, with their convexity downwards, concentrically surrounding with a rim imperfectly formed convolutions. The fissura præcentralis goes as far as the median surface, and contains an operculum in the transition between the second and third frontal convolutions. The same is found (between all the three frontal convolutions) in the most anterior part of the first frontal furrow. The second frontal convolution at the base badly developed; the fissura cruciata showing a complicated operculum. The gyri hippocampi and uncinatus strongly projecting over the frontal lobes and a little furrowed. The gyri fusiformis and lingualis rising steeply towards the top of the occiput, much diminished, and especially the first very much shortened by the first two temporal convolutions, and connected in a peculiar serpentine manner with the posterior combining part of the lobulus tuberis and the second temporal lobe at the parieto temporal operculum. The perpendicular occipital fissure is connected with the sulcus hippocampi, and conceals numerous *plis de passage.* Especially the median part of the occipital lobe reduced to a minimum. On the left, the posterior ascending spur of the fissure of the fossa Sylvii and the interparietal fissure exhibit their normal proportion, but the latter conceals in it numerous *plis de passage.* The horizontal occipital fissure exhibits the apestructure. The parieto-temporal fissure is not well marked, but the parieto-temporal operculum is. The vertical occipital fissure is connected with the sulcus hippocampi. The first temporal furrow shows the brute form by a posterior spur (the concavity directed downwards). The fissura Rolandi is connected anteriorly with the first and third frontal furrows, and in all the fissures there are numerous opercular structures. The third frontal convolution is submerged at the base.

In the case of the first murderer's brain examined, the space of the *plis de passage* was, in the part behind the posterior central convolution, almost predominant over the developed convolutions, so that a normal type of convolutions and furrows could scarcely be drawn. Here, also, the furrow of Rolandi communicated with the fissura præcentralis. The occipital lobes scarcely covered the cerebellum.

For other writings on this subject the reader is referred to works of Leuret, Gratiolet, and Broca, in France; Owen and Huxley, and their school, in England; Huschke, Virchow, and Bischoff, in Germany, and Lombroso, in Italy.

CHAPTER XIV.

The Psychological Aspects of the Trial of Edward Newton Rowell, at Batavia, New York.

On the 30th of October, 1883, E. Newton Rowell, of Batavia, N. Y., shot and killed Johnson Lynch, of Utica, N. Y., in the act of adultery. The circumstances of the case as developed in the trial before the accomplished Supreme Court, Judge Haight, of Genesee county, render it both very interesting and unique. Since the trial of Daniel E. Sickles, in 1859, and of Daniel Macfarland, in 1870, there has hardly been so interesting a case.

Rowell's brain had long been in that condition where it was only necessary for some sudden and violent emotion, related to the main cause (the revelations concerning his wife that his partner, Palmer, kept telling him about), to act upon a congested brain and precipitate him into the manslaughter which he committed. Having suffered from melancholia for a long time, two years, arising primarily from business perplexities, and added to by the dwelling of his mind on the gradual alienation of affection on the part of Mrs. Rowell from himself; having inherited a neuropathic, unstable nervous organization from his ancestors, there being consumption, epilepsy, insanity and inebriety in the family record, we have here a man predisposed to the acquisition of insanity, on even slight exciting causes (and he had a great cause), and who had not an organization fitting him to cope successfully with even the ordinary disappointments and calamities of life. The principal exciting

cause of Rowell's mental disturbance had been operating
for some time, and being a well developed case of melancholia, having had epileptic attacks at intervals also, is
it at all strange that, brought into contact with the man
who had been the chief exciting cause of his domestic
trouble and grief, which in its turn had added great
intensity to incipient melancholia, and brought into contact with him in his own house, seducing his own wife,
that great irritation of the brain, amounting to frenzy
or a maniacal outbust, ensued, during which Rowell shot
and killed Lynch.   These are scientific facts bearing on
Rowell's mental disorder, which was unquestionable,
gathered from the mass of testimony of many witnesses,
who testified to the extremely excited mental condition
in which they had all seen him previous to the shooting,
and all the circumstances of the case negative the pos-
sibity of Rowell's mental condition being one of sedate-
ness or coolness antecedent to the shooting.   The impression and opinion of all who met Rowell between August
20th and October 30th, 1883, was that he was not in his
right mind ; and at the writer's examination of him at
the Batavia jail, November 15th, 1883, which will be
presently detailed, he told the writer that there had been
many times, particularly within these dates, when he
seriously meditated ending his existence by his own
hand.   He did not wish to live.   There was not the
slightest attempt on the part of the defense to conjure
up insanity as a defense for a crime which any jury
would consider justifiable, on account of the greatness of
the provocation and of Rowell's previous good character.
Rowell was, furthermore, in his own house — his castle ;
and although he was physically so disproportioned to the
physical power of Lynch, he would have been a poltroon
indeed, if ever so sane a man, if he would not shoot any
man down who dare dishonor his wife's person in his
presence.   Rowell does not need the justification of
the plea of insanity in the eyes of the world.   An
undoubted case of melancholia, a condition in which

he had meditated suicide, and with insanity, epilepsy and consumption in his ancestry, he was, by Lynch's presence, thrown into a state of excitement in which he was divested of his reason and judgment, and was deprived of his mental power to an extent placing him beyond the range of self-control, so that he had no will power to restrain himself from the commission of the act alleged against him at the time of its commission, and was entitled to an acquittal as "not guilty, on the ground of insanity." The jury have been criticised for their verdict of "not guilty, on the ground of self-defense," but unjustly, we think, as they did so on law points which were contained in the judge's charge, which was calm, well-conceived and impartial, and also eloquent. They, however, all concurred in Rowell's mental unsoundness. Rowell had had no will or intellectual power for some time antecedent, and, according to the testimony, had been a mere machine in the hands of his partner, who really planned the circumstance by which Rowell went to the hotel first, and then subsequently secreted himself in a closet in his own house, carrying with him such ill-assorted, incongruous weapons as no sane man of Rowell's previous conduct and character would ever have thought of, and which were suggested to him by his partner. The mental pabulum which had fed his partner's brain cells about this time must have been evolved from the perusal of sensational dime novel literature, as a quantity of pepper to blind the eyes, a sling-shot made by putting a stone in the heel of a stocking and a revolver were prescribed by him to Rowell as about the correct armamentarium for him in case he were to be attacked in his own house. Rowell, in his mental automatism, accepted these suggestions, and did as he was told. The design of Rowell's partner was to make up a party who should find Lynch and turn him, disgraced, into the street. From the evidence given by gentlemen from Utica of high standing, it is utterly inconceivable that such a man as they describe Rowell

to be while a citizen of Utica, a calm, intelligent, gentle-
manly man, of good standing and character, could lend
himself to or abet any such proceeding, or go about it in
any such manner as Rowell did.   It showed most
markedly the degradation of his intellect that he did so.
Even the evidence of the prosecution in this case aided
very materially the evidence established by the defense,
of a state of facts consistent with the prisoner's insane
state of mind previous to the shooting.

There had been sleeplessness for some years, during
which time only four or five hours of sleep nightly had
been obtained, and thus the nutritional disturbance of
the brain had gone on ; the psychic signs of the neuras-
thenia preceding insanity had all shown themselves.
Within three years there had been a marked change in
character and disposition.   Prior to that time Rowell
was clear-headed and calm and persistent.   Latterly, he
had tired easily, and was not capable of doing, either
mentally or physically, what he had easily accomplished
before this.   He would change easily, and there was a
peculiar suddenness in determination, with no deliberation, quite foreign to him when in his normal state.
He did not act upon mature reflection.   He removed
from Utica, suddenly, and under great excitement, to
Batavia, the scene of the tragedy.   He became very
emotional.   He tendered to a certain person security for
an obligation not incurred.   He suffered from epileptic
vertigo, and this would come on as the result of any
mental or physical exertion.   He had completely lost
his former mental and physical power.   On the night of
the homicide, within three minutes afterwards, he was
found pale and talked rapidly and excitedly.   He paid
no attention whatever, as a sane man would have done,
to the caution given him not to talk, but was even child-
ishly garrulous and went on heedlessly.   For some time
it had been remarked that even when he was laboring
under slight excitement the extremities became cold and
the head hot, and that this condition would last until the·

circulation reëstablished itself. At other times he would be unduly hot, and on the coldest nights would throw off all the clothes ; had complained of sensations in his head as if some one had struck him a blow on the head. There had been also great difficulty in holding his attention on one subject for any length of time, so that more than one citizen of Batavia had openly expressed themselves that Rowell was "a lunatic or a fool." Normally, he could hold his attention on any given subject for *any* length of time. Many witnesses detailed rambling conversations with some incoherence, before the shooting, and his melancholic state of mind was patent to all who came in contact with him. He would sit in a corner motionless for hours, with his head in his hands, and was constipated and sleepless.

November 15th, 1893, by request of the counsel for the defense, the writer visited Rowell in jail, spending three hours with him. In attitude, dress, conversation and gesture he was a typical melancholiac. In talking, he viewed everything through a dark, gloomy medium, and seemed rather sorry that he had not carried out his impulses to suicide, which he talked freely about, saying that only his love for his children had kept him back. His head was hot; the extremities were cold ; the pulse 90; the tongue was coated with a thick fur and dry ; the breath very foul; the eyes injected and suffused ; the skin was harsh and dry, with no normal elasticity; the hair was dry and harsh ; the complexion muddy and unwholesome ; when he moved, it was with that peculiar abruptness characteristic of the insane. There was great constipation, from his account, and great sleeplessness. There was unequal muscular power of the two sides of the body, and a peculiar droop of the shoulders when walking. He did not hold himself erect, but looked and acted a sick man, mentally and physically. There was dullness at the apices of both lungs ; the heart sounds were normal ; the urine scanty and high colored.

As the mind did not appear to be preoccupied with

any delusive idea, or by maniacal excitement, the diagnosis was a case of simple melancholia with suicidal and
homicidal tendencies, complicated with epilepsy. The
facial expression was one of deep and concentrated
misery and unhappiness. Rowell was very ready to
converse on his mental symptoms. He was thin and
emaciated, owing to loss and wasting of the tissues from
increased activity of the mental functions and derangement of the alimentative processes. There were also
signs of gastric and hepatic disorder observable in the
tongue, the skin and the nutrition of the body. His
postures and gestures exhibited perfectly and diagnos-
tically the mental condition of melancholia under which
he was laboring.

The attitudes were fixed and the gestures slow. The
eyebrows drooped, and the prevailing emotion of melancholy had stamped itself on his physiognomy. His
whole appearance, conduct and conversation were those
of a decided case of melancholia. In testing his memory
and attention it was found that while his memory for
past events was good, his memory for recent events was
very poor indeed. It was very difficult for him to
pursue consecutively any train of thought. A careful
conversation revealed no delusions, either respecting his
powers of body and mind, in relation to persons, or with
reference to any subject or object whatever. There were
no perverted emotions, no insane delusions respecting
religion. He stood the test of careful inquiry on all
subjects well, except as regarded his domestic trouble and
grief, when he became quite excited. He had the sad
and anxious eye, the drooping brow, the attenuated and
careworn features, the muddy complexion and harsh
skin, the inertia of body, the stooping and crouching
posture, characteristic of great gloom and wretchedness,
and if he had never committed any overt act, and had
been brought to my office, I should have had no hesitation in pronouncing the patient insane, and should
have expressed my fear of the possibilities of

suicide, and ordered the immediate surveillance of an intelligent nurse, to be with him night and day, together with appropriate medical treatment.

On trial, Miss Julia Rowell, A. Gage, an insurance agent, Charles Wescott, Rose Sargeant, Joseph C. Wilson, a capitalist of Batavia, Isaac Garson, Julia Fish, and many others, all testified that Rowell had exhibited for some months pallor and strangeness of manner; that he looked worn and unnatural in body, face and eyes; that he was thin and emaciated; that he had a peculiar glare and staring eyes, and that his conduct had impressed them all for some time as unnatural and irrational; that a marked change in charactar and disposition had manifested itself in him. Even the witnesses called by the prosecution testified substantially to these same facts, on cross-examination. Several eminent gentlemen from Utica testified as to Rowell's irreproachable character while in Utica, both in early and in later life, among others, Andrew McMillan, Esq., and Charles W. Hackett, Esq.

Mrs. Sophia G. Balcom, one of the witnesses, of Marshalltown, Iowa, wife of a physician, testified by affirmation: Am Mr. Rowell's mother's sister; there were four sisters and one brother; Rowell's mother died when he was eight years old; I lived in Utica then; she was next to the oldest; think she was twenty-nine or thirty; her constitution was feeble; she was despondent, and at times excitable and nervous before Rowell's birth; her lungs were diseased for years; my brother had four girls and two boys; two of her sisters were insane; I saw both for years; Eunice Shear was insane when forty-five, and died insane; the sisters were kept under constant watch in the house; they were not capable of caring for themselves; they were moody, dejected and violent at times; the youngest, Nancy, was insane at nineteen, and lived between forty and fifty years, and died insane; my mother's father was insane; his name was Begun.

Cross-examination: To the District Attorney — Left

New York State sixteen years ago; have lived two years in Marshalltown; left Utica when I was twenty and unmarried; lived in Hudson and Syracuse; all of my family dead; Alvira was Rowell's mother; he was born in Utica; she was small and slender; he was delicate from infancy; she had three children; all are in court; do not know the cause of Nancy Begun's insanity; would have to write out a book on what I had heard of it [laughter]; did not hear that insanity was incident to female disease; my impression was that it was inherited; my brother was in the South at the war; do not think him alive.

ROWELL'S FATHER.

Edward Rowell, father of the accused, testified: Am a machinist, of Utica; have been in the employ of Philo S. Curtis twenty-one years; my father was not rugged, and was given up by the doctors early, from consumption; he was sick and nervous; he died when sixty-eight; my mother was of a nervous temperament; she died when about fifty; my parents came from England in 1833; they brought over seven children, including myself; there was lung disease and consumption in father's family; I am not strong; from twenty-one to thirty-one had dizziness and vertigo, which produced cloudiness of vision for a few years; fell unconscious in the shop at New York Mills in 1842 or 1843; had no warning of the attack; after my first fall and unconsciousness I was confused and weak; the second attack was while I was shoveling snow; I fell on the walk; had another attack recently; have always been rather weak; have had neuralgia in the forehead and face eighteen or twenty years so as to disable me; my son's mother was nervous and very much inclined to be melancholy; think she died at thirty-one; his mother's father drank to excess; never saw him sober; my son Newton was always weak physically; his mind was bright and active; his mind was much in advance of his body; he had lung trouble and cough, and was threatened with consumption ; he suffered a great deal from severe headache in the forehead, he has been for twelve or fifteen years sleepless at times ; any particular excitement or trouble made him sleepless ; he was subject to nightmare ; in 1871 or 1872 he fell from a step-ladder in Newell's store and was brought home ; he complained of his head for a week or two.

The writer was put on the witness stand as the expert for the defense, and his evidence borne out and followed by Dr. S. T. Clark, of Lockport, N. Y., Dr. Herbert A. Morse, of Batavia, and Dr. M. W. Townsend, of Bergen, all of them gentlemen of ability and high standing in the medical profession. The expert testimony elicited is as follows, although in a very condensed form :

After a few minutes of waiting, Edward C. Mann was called and said : I reside in New York ; I am a physician ; my specialty is nervous and mental diseases ; have practiced thirteen years, and have had a specialty for ten years ; I have a private hospital ; I am a member of the New York County Medical Society ; I examined the defendant in the jail at Batavia in November ; I made a physical examination first and questioned him about his health ; his face was emaciated, and its expression gloomy and dejected, eyes suffused, complexion muddy, skin harsh and dry ; when I requested him to move, his motions were abnormally quick and nervous ; his muscles were flabby and had lost their tenacity ; there was a dullness at the apex of the lungs ; it would indicate a predisposition to consumption ; his cold feet and hands indicated poor circulation ; his tongue was coated and breath foul ; his pulse was 90 ; there was loss of power on his right side. In conversing with him I found him more excitable than a normal man should be. The nutrition was poor. It was a decided case of melancholia, one of the forms of mental disease. Insanity has never been satisfactorily defined in all respects. Insanity is a disease of the body, affecting the mind in such a way as to render an individual irre-

sponsible. All lunatics are not raving maniacs. Sometimes there is no raving at all. The organ of thought is
the brain. The brain is wasted by work, but is constantly
repaired by nature by food and sleep. Ill health is
a frequent cause of insanity, as is domestic unhappiness and grief. Sleeplessness accompanies insanity.
Insanity is remarkably hereditary. A sudden shock
is a cause in a person liable to insanity. Epilepsy is
divided into great and . lesser. Continued epileptic
seizures affect the mind and weaken it, rendering the
brain sensitive to excitement. The symptoms of insanity
may be mental depression, exaltation, or weakness. Exaltation would be shown in mania. Depression would be
shown in a gloomy, melancholy person. Weakness
is seen in idiocy and dementia. When a person
becomes insane there is a change in his manner ; would
lose interest in business, would be irritable, gloomy, and,
when under arrest, after an overt act, would not seek to
escape. A sane man might run away from the officer,
while an insane one does not. Emotional display, characteristic postures and attitudes, sleeplessness, constipation, gloom and depression, harshness of the skin, inability to concentrate attention, impairment of memory
and judgment, are signs of melancholia ; other symptoms
are high or slow pulse and gloominess. Melancholia is
one of the most dangerous phases of insanity. Rash acts
are liable to be committed at any stage of the disease,
which is accompanied by loss of will ; the sight of a
weapon suggests its use. The insane plan and design as
the sane do ; the distinguishing feature is the loss of
self-control. Paroxysms of violence occur in melancholia ; they can be roused to an uncontrollable fury,
followed by a calm.

At 2.30 P. M. Judge Sutton read the following hypothetical question :

Q. Suppose an individual of high and unblemished
character ; of bilious, nervous temperament ; descended
from an enfeebled parentage ; begotten of a race in which

insanity was hereditary ; of diurnal and frequent nocturnal epileptiform seizures, occurring in the early morning ;
who, after months of intense mental agony and emotional
disturbance, sleepless nights, struggling with alternate
hope and fear concerning the chastity of his wife, to whom
he is devotedly attached, has lost his interest in business ;
his judgment and memory have become impaired ; has
become dejected and changed, melancholy, morose and
irritable ; who, when met on the streets, was seen to present a wild and staring gaze, failing to give ordinary
recognition to conversation, acting in an unnatural and
irrational manner, incoherent ; who, from an independent, executive person, becomes negative, surrendering
his own personal resistance to an officious friend, accepting his advice in place of his own judgment, at whose
instigation he arms himself with a pound of powdered
pepper, a stocking, in the toe of which has been placed
a stone, and a revolving pistol ; who, at the direction of
that adviser, repairs to his own house for the purpose of
obtaining evidence which shall prove or disprove his
wife's faithfulness, and there, finding his wife in adul-
terous connection with a man, shoots and kills the man ;
who immediately after the killing was calm and absent-
minded ; who made no attempt to escape the officers or
to deny the killing ; who on the way to the magistrate's
office was silent, motionless, careless, unconcerned and
unmoved in the midst of intense excitement ; whose
physical condition was emaciated and weak, complexion
pale and sallow, motions quick and nervous, and who for
two days prior to the shooting had taken little nourishment. Under all these circumstances did the individual,
at the time of shooting, labor under such a defect of
reason as not to know the nature and quality of the act
he was doing, and not to know that the act was wrong ?

Mr. North objected to the question, claiming that the
suppositions were not correct, as shown by the evidence.
"There is no evidence as to nocturnal or diurnal epileptic seizures, sleepless nights, loss of memory, or that

he was armed with a pistol at the advice of an officious friend."

A. Assuming the truth of all the suppositions in that question, the man was insane.

Cross-examined by Mr. North: I was there from 2 to 5 P. M.; his brother was there; the prisoner did not know who I was; he answered all my questions; I asked nothing about the shooting; an insane man moves as no other man does; it is rare to find anything uniform in mental diseases; persons sometimes recover from melancholia in three months; melancholia induced by epilepsy would be accompanied by destructive impulses; insane people often remember their acts and can recount them when sane; unconsciousness is characteristic of the epileptic paroxysm; the occurrences of violence are instantaneous, and occur on the impulse of the moment; their acts are sometimes performed with an apparent motive and sometimes not; if there was a motive he would respond to it, and if not he would respond to the disease; I have weapons that lunatics have spent months in making, for destructive purposes; heredity is an important element in insanity; insanity, as a rule, skips a generation; an insane grandfather is more dangerous than an insane father.

Q. What do you think of an insane great-grandfather?

A. That fact has not been testified to.

Insanity in collateral relatives is a strong evidence of taint in the family. A man's recovery who was insane, as in the hypothetical question, would depend on his rest and tranquility. From three to six months would be about the usual time, under fairly favorable circumstances. The best treatment for melancholia is food, rest and quiet, with warm baths, sedatives and tonics. Have known lunatics to commit a crime, saying that they were insane and could not be punished. Many lunatics know the difference between right and wrong.

Q. Take a man of thirty-six, who, in boyhood, was an attentive pupil; had early in life gone into business; intent on all he had to do; continued so; married at the age of twenty-nine; had two children; was living with his family; about August, in a given year, was informed that his wife had committed adultery with a man in another town; talked about the case, and consulted a lawyer, who said there was not evidence enough to secure a divorce; talks over the way to procure a divorce; lays a plan to pretend to go away from town; leaves his home on Monday; goes to an obscure hotel to watch trains; sees his friend and partner often; late in second day receives intelligence that the adulterer is coming; puts pepper in his pocket; takes a pistol and stocking; goes to the house and secretes himself till the guest's arrival; hears the conversation; waits till his wife and her guest come to bed; feels over them in the bed and fires his pistol at them, and the man starting, he fires several times; goes down stair and cries, "Help, for God's sake, help, I've killed a man;" gets a basin of water and calls a physician; he stands by him, raises his hand, asks a doctor to go to his wife; endeavors to overrule her; says to her, "Jennie, I have warned you of these things;" goes to the justice's office; relates to the officers all about his crime on the way; sits quietly in the office during the testimony and pleads not guilty; is taken to jail; relates the facts to a friend who calls on him in the jail next day. Is such a man sane or insane?

A. I should say, in view of all the evidence on this case, you had detailed the facts concerning an insane man.

A murmur of applause was raised in the audience, but promptly put down.

The Court: Suppose a man advised another to use a slungshot made from a stocking.

The Witness: It would be a very weak man that took such advice.

The Court: Let me add to the people's question.

Assuming the facts as given, did the man at the time of the act know or not if he was doing wrong?

A. That would depend on the circumstances. He might and he might not. His taking the number of weapons is an evidence of insanity.

Judge Sutton : Take the District Attorney's question, and fill it out with depression, excitement, gloominess, and all the facts I have given you of his appearance ; would he be sane or insane ?

Mr. North : I object.

The Court : The defense has made its question and the people have made theirs. The jury can judge which of the two conforms to the facts in this case. Let him answer the question.

Judge Sutton : Read it, Mr. Stenographer. I want to add a little more.

Emaciation and loss in business were added, and Dr. Mann said the man was insane.

Q. (By Judge Sutton) Suppose a man, working in a paper-hanging establishment, fell from a step-ladder, without any apparent cause, what would you consider it ?

A. I should say it might be an attack of vertigo.

Q. If you knew he had epileptic seizures at night, what then?

A. It would surely be an attack of epilepsy.

Judge Sutton detailed the attack that Wm. Rowell had when he came in one morning from milking, as testified to a day or two since, and asked what that was, and the witness characterized it as epilepsy. Other things that happened to Wm. Rowell were gone over, and they were designated as epilepsy.

Mr. North objected, saying it was not in the evidence.

Judge Sutton : Oh, I am within the evidence, unless I have had a seizure.

The Court : It is not important. Let him answer.

The case of Mr. Edward Rowell was gone over, and designated as a nervous disease or epilepsy.

The Witness : It is not unusual for a person with melancholia to brighten up when a visitor comes in. A man attended to as Rowell was, would have mental tranquility in place of mental agitation, though a jail is not the best place to cure a man. A man with epilepsy can have an unconscious minute and never know it. Dr. Clouston is an eminent authority on mental diseases; he says three months is a usual time for recovery in mild cases.

(To Mr. North) The circumstances surrounding Rowell were not at all favorable to recovery.

The Court: Would the fact that he was resting under a criminal charge affect his recovery?

The Witness: A man mentally diseased don't look at such things as a sane man does; in jail he would have rest and quiet.

There is no improbability in epilepsy and melancholia existing together; increased flesh would be an evidence of improvement. At 3.55 Dr. Mann left the stand, and Dr. Simeon P. Clark was sworn.

Dr. Clark said: I am a physician at Lockport; I have practiced since 1860; I have paid attention to diseases of the brain; have written for journals on the subject; Bucknill and Tuke are good authority on mental diseases; some of my language is incorporated in their work.

The counsel got into a little discussion, and were hurried on by the Court's impatient "Hasten on, gentlemen."

The Witness: I have had experience in the treatment of the insane.

The hypothetical question was read to Dr. Clark, and he answered that the person described was "assuredly insane."

Domestic grief and ill health are a frequent cause of insanity; insanity and epilepsy, in common with all nervous diseases, are hereditary. Emaciation, change, sleeplessness, irritability, or lack of it, moodiness, gloom, surrender to justice on commission of a crime, dry, harsh skin, constipation, muscular restlessness, emotional excitability, impairment of memory, coated tongue and high pulse are symptoms of insanity. Melancholia is sometimes attended with very violent acts, and often that is a symptom. A weapon is suggestive of an act of violence. The insane frequently have motives.

Cross-examination : In 1867 I had charge of the Niagara county insane, of whom there were twenty-five or thirty ; I was the county physician.

The District Attorney cross-examined Dr. Clark in much the same line gone over by him in questioning Dr. Mann. The doctors disagreed on some minor points. When the District Attorney's hypothetical question was read, Dr. Clark said : "I think he might be sane or insane ; the difference would depend on the condition of the man. You have given almost data enough to make melancholia here.

Dr. Herbert A. Morse, of Batavia, was the next witness. Dr. Morse said he had not made a specialty of diseases of the mind. Miss Julia Rowell, of Utica, within half an hour, has had an epileptiform seizure, and was under my care.

To the hypothetical question, Dr. Morse said : I should think he was. I should call him insane."

"You may examine," said Judge Sutton, when this question was answered.

Cross-examination by Mr. North : Miss Rowell walked to the door of the Court House, was placed on a couch ; she had convulsive motions of the arms and face ; was unconscious for a minute ; she had an epileptiform attack.

In answer to the people's hypothetical question, Dr. Morse said : "The person described in that question, to my mind, bears slight evidence to insanity."

Dr. M. W. Townsend, of Bergen, said : I am a physician and surgeon ; mental diseases have been part of my study.

To the hypothetical question he said, "I think, if he was in the condition spoken of, he was insane."

The District Attorney cross-examined the doctor, who

said: "It is impossible to rouse a person from an epileptic fit." To the people's hypothetical question he said: "I could not state exactly; I see one evidence of a mental delinquent in the question; I refer to his taking multitudinous weapons."

To Judge Sutton the witness said that Miss Julia Rowell did not arouse her brother from his epileptic seizures.

The writer defined insanity as "*a disease of the body affecting the mind by deranging its faculties, and causing such a suspension or impairment of the healthy intellect, emotions, or the will, as to render an individual irresponsible.*" Also, that in insanity *there was disease of the body, affecting the mind, causing loss of self-control.*

Dr. Judson B. Andrews, the accomplished superintendent of the Buffalo Insane Asylum, was the expert witness for the prosecution, and testified substantially as follows:

Have been in the practice of my profession and in the care of the insane about seventeen years; I have 340 patients in Buffalo, and had 600 at Utica; have been in court since Thursday; have heard the hypothetical question of the people; I should say that the individual characterized in that question was sane.

Cross-examination by Judge Sutton: That would be my opinion; I suppose that he might be insane.

Q. Would you be influenced in your opinion by other details prior to, and in other matters subsequent to the shooting?

A. I cannot answer that yes or no.

The Court: You may ask the question in another way.

Q. You say he might be insane?

A. Not without there were other elements in the case; there might be other evidence that would satisfy medical experts that he was insane.

Judge Sutton propounded his hypothetical question.

Witness: That man was insane. [Applause in the court room].

Witness : Insanity is a disease of the nervous mass ; I mean that the mass within the skull must be affected before insanity results ; the common idea that insane persons are always raving and incoherent is a mistaken idea; a great proportion are neither raving, incoherent nor wild ; ordinarily a large class are as well behaved as people outside of the asylum ; the brain substance is destroyed in use and repaired from blood, which is made of the food and air taken into the body and lungs ; if blood-making is interrupted, the brain will suffer from lack of nutrition ; it might impair the mental faculties ; the form of disturbance may be different in some persons than in others ; some insane people reason well ; if they did not reason from delusions in some cases it would be difficult to declare them insane ; ill-health is largely the cause of insanity ; domestic grief, afflictions and sleeplessness are great causes of insanity in affectable persons ; the mind dwelling upon any grief or affliction would disturb the mental equilibrium ; if continued, the brain will break down ; if the equilibrium is disturbed, insanity may follow ; epilepsy is a disease of the brain ; if epilepsy was added to the facts stated in the question, it would have enough to build insanity upon ; the ordinary result of epilepsy tends to mental deterioration ; great mental strain upon an epileptic may produce insanity ; sleeplessness is a cause and symptom of insanity ; also emaciation and change in temperament ; melancholia or mania may follow ; no one can tell which way a patient will go ; cases differ, as do men's faces ; irritability is a symptom and accompaniment of insanity, as are also moodiness and gloominess; some cases are given in which insane persons have surrendered to the officers after commission of crime.

Judge Sutton read from a paper delivered by Dr. Andrews, on premonitory symptoms of insanity, before the Alumni of the Buffalo Medical College. Of course,

the doctor admitted the correctness of the theories
included in the paper. The theories seemed to support
the claim of the defense, that Rowell was insane, and
many of them covered exactly many points in defense
upon which the counsel has laid heavy stress. There were
many pleasantries between counsel and Dr. Andrews, who
bore himself in his usual intelligent, clear, dignified and
good-natured manner.

The summing up of the defense by Wm. C. Watson,
Esq., of Batavia, was an eloquent effort.

JUDGE HAIGHT'S CHARGE.

At 3.55 p. m. the jurors stood up in their places, and
Judge Haight opened his charge, which was as follows:

Gentlemen of the jury: You are impaneled to try the
defendant, who stands charged with the crime of manslaughter in the first degree, in killing Johnson L. Lynch,
on the 30th of October last. It now becomes your duty,
from the evidence, to determine his guilt or innocence.
Each of you, gentlemen before entering the jury box,
had read and heard something of the accounts of this
affair, from which you had formed some impression in
reference to the transactions. Each of you, however,
stated on your oath that you believed that you could lay
aside your previously formed opinion; you could enter
the jury box and listen to the evidence and determine
the facts anew, according to law and evidence, without
being influenced by any previously formed opinion. This
duty now, gentlemen, devolves upon you. It is a duty
you owe to the public. You owe it to the prisoner. You
owe it to your own consciences. It is not the judgment
of any one of your number we seek. It is the united
judgment and wisdom of your entire body. Let me,
therefore, caution you, in going to your jury room, not
to enter into any hasty or passionate discussion of the
questions involved, but to coolly, calmly reason one with
another, to the end, if possible, that you may bring your
minds to a common conclusion, and in doing so determine right in this matter. Something has been said,

gentlemen, in reference to our laws — in regard to that of murder and that of adultery. Some criticism has passed the lips of counsel in reference to them, and we have had our attention called to the laws of other countries. It is possible, gentlemen, that our laws are defective upon the subject of adultery. I am not prepared to say but they are. But we are not the law-makers. We are here sworn to execute the law. On one occasion, a frail, trembling woman, guilty of adultery, was led by an excited crowd before Christ. The penalty of the crime was death by stoning. He commanded that he who was without sin should cast the first stone. No stone being cast, she was told to go her way in peace and sin no more. From that day on it became a serious question in the minds of Christian people as to what punishment ought to be meted out for this offense, until finally, in the State of New York, in various States of the Union, and in Christian Europe, it has come to be regarded as a sin which God, in his own way and in his own good time, would probably punish. But, gentlemen, we are not here to try this question; only so far as the act operates upon the mind of the prisoner is it properly before us for our consideration. The question that we are here to try is the killing of Lynch, and that we are to determine from the well-settled law of our land. In your deliberation, gentlemen, upon this question it becomes necessary for you to understand something in reference to the rules of evidence. Evidence is of two kinds, and commonly known as direct and circumstantial. Direct evidence is that which is sworn to by a witness upon the stand from knowledge that he has derived from his own senses, as to where he has seen and heard that which he relates. Circumstantial evidence, when collected together, convinces the mind as to what the facts are. It is commonly likened to a chain with numerous links. Again, I have heard it likened to a goblet broken into pieces. The broken pieces, when gathered together, match one upon another, and satisfy the mind that they once formed an entire goblet. And so it is with circumstantial evidence. Again, gentlemen, it is the rule of criminal evidence that before a jury convicts a person charged with the crime, the jury must become satisfied of the guilt of the accused beyond a reasonable doubt, and if there still exists in the minds of the jury a reasonable doubt in reference to any of the propositions it becomes necessary for the people to establish, it would become necessary for the jury to give the prisoner the benefit of such doubt. So that, in case you have reasonable doubt as to his guilt as to the killing, or as to whether he did it in self-defense, or if you still have reasonable doubt in reference to his being sane at the time of the killing, then, as I have stated before, the law steps in and says such doubts will be given to the prisoner. What is a reasonable doubt, gentlemen?—for, in order to enable you to determine to what degree you must be satisfied, you must understand what is meant by a reasonable doubt. It is not every conjecture of the human mind. Reasonable doubt, in order to make it a reasonable doubt, must be founded on some fact, some circumstance or theory which would create reasonable doubt in the ordinary mind of man. It must be founded on some fact, upon some theory or circumstance which would create such a doubt. The next question, gentlemen, it becomes necessary for us to consider is the law pertaining to homicide. The law upon this subject has been framed with great care. It has been subjected to much discussion and deliberation on the part of our able jurors and legislators. Every sentence in it has been discussed and deliberated upon. The wisdom and experience of all the civilized countries of the globe have been considered in framing its provisions. Still, it may be imperfect. Few things that a man does are absolutely perfect, but in the main I believe the laws to be wise and just. It becomes our duty to ascertain their provision as they apply to the evidence to which we have listened, and then to determine whether or not the prisoner is guilty of the crime charged against him.

The first provision of the statute to which I call your attention is in reference to homicide, excusable and justifiable. Homicide is excusable in an individual when it is committed by accident and by misfortune in the doing of lawful acts, and with ordinary caution — as, for instance, suppose you were engaged in chopping, and you are standing upon a log, if you please, exercising ordinary care and caution; but by reason of some ice that has escaped your attention, a foot slips, you lose your balance, the axe falls from your hand, and in doing so strikes and kills another. It is purely accidental. It would be excusable. Again, it is justifiable, when performed by an officer in the discharge of his duty, and in obedience to the judgment of a court having jurisdiction to pronounce the punishment of death. It is justifiable on the part of an individual in actual resistance of an attempt to commit felony upon the slayer in his presence or in the dwelling, or other place of abode in which he is. If a person should find a burglar in his house in the night time, in case you should kill him under such circumstances, the law would then step in and justify you; and this is upon the theory that the burglar who enters the house of an individual in the night time, for the purpose of committing a crime, goes armed and prepared to protect himself by killing another, in case it becomes necessary, in order to avoid being taken himself. There are various cases in which it is justifiable, but which are not applicable and necessary for you to take into consideration, except that which is embraced in the section of the statute, to which I now call your attention especially. Homicide is also justifiable when committed either in lawful defense of the slayer, or of his or her husband, wife, parent, child, brother, sister, master or servant, or of any other person in his presence or company, when there is reasonable ground to apprehend a design on the part of the person slain to commit a felony, or to do some great personal injury to the slayer, or to any such person, and there is

imminent danger of such design being accomplished.   So
that, gentlemen, under this provision of the statute, in
case a person makes an assault or attack upon you, under
such circumstances as to lead a reasonable man to believe
that he is about to kill or to do you great bodily injury,
and there is imminent danger of his doing so, then the
person has the right to kill, and he would be justified in
the act; and that is the right a husband has to extend
toward his wife in case an assault is being made upon
her under such circumstances so that she is liable to
suffer great bodily injury, or to have her life taken by
the assailant.   A father may exercise this power, under
the statute, in the protection of the lives of his children
and his servant and his sister.   In case a husband should
find a person committing an assault upon the person of
his wife, attempting to force her to have carnal connection with him against her will and resistance — if he kill
a person engaged in making such an assault, he would
be justified under the statute.   In case an assault of this
character was being made upon his daughter or upon his
sister, he would have the right to protect the person from
such an assault by killing the person who was making
the assault.   But, gentlemen, where the assault is made
with the consent of the woman, where they are having a
mutual intercourse, then the law does not justify the
taking of human life.

The next provision of the statute to which I call your
attention is that of murder in the first degree.   The
killing of a human being, unless it is excusable or justifiable, is murder in the first degree, when committed
from deliberate and premature design to effect the death
of the person killed.   I read only that portion which
becomes necessary to consider in the further determination of this case.   Under the provisions of the statute
defining murder in the first degree, there must be a
deliberation and premeditated design.   The life of a
person must be taken from deliberate and premeditated
design to effect the death of the person killed.   It is

claimed, on the part of the prosecution, in the case under consideration, that there may have been deliberation and premeditation during the time intervening between August 17th and October 30th. We still may believe that the deliberation and premeditation was not to take life, but to obtain some evidence and inflict some other punishment than that of the taking of life. It is claimed that the defense under consideration was not that of murder in the second degree. Such killing of a human being is murder in the second degree when committed in a design to effect the death of the person killed, or of another, but without deliberation and premeditation. The difference is that in murder of the first degree there must be previous deliberation and premeditation. In second degree, deliberation and premeditation is no part of the crime, but it must be the intentional killing. It is claimed, on the part of the prosecution, that the intention of this defendant in going to the house, under the circumstances in which this affray took place, was not to kill, but to otherwise punish, and that he is not guilty of murder in the second degree. The next offense under the statute is that of manslaughter in the first degree. Such homicide as manslaughter in the first degree is when committed without design to effect death, and when in the heat of passion, but in a cruel and unusual manner, or by means of a dangerous weapon. The attorney claims that the prisoner is guilty of manslaughter in the second degree. It varies from that in the first degree in that the killing is not required to be by means of a dangerous weapon.

So much, gentlemen, for law in this case. It becomes your duty, from the evidence to which you have listened, to determine, in the first place, whether or not the crime of manslaughter has been committed by this prisoner. The evidence we have bearing upon this subject is the testimony of his neighbor Swanson, who lived the next door to the defendant. He testifies that on the occasion of the shooting he was sitting by his window, and first

heard a pistol shot, and some four or five seconds thereafter he thinks he heard another, much louder, and he
tells us he thinks he heard two other shots fired in quick
succession.   He went to the door.   Defendant came out
upon the piazza and cried, "Help! Help! For God's
sake come over.   I have shot a man."   We are told
Swanson returned to his room and put on his shoes, and
got his neighbor, Mr. Reed, and they went over.   On
entering the hallway, they entered into conversation
with the prisoner.   Inquiry was made of him if this man
was a burglar in the house, and his answer was "No,"
that he found him seducing his wife, and had shot him.
This was in substance the conversation that took place
between them in reference to the act of killing.   There
was other conversation, but it is not necessary for me
to refer to it in detail, for it is only in reference to
the killing, and as to the mattter of the killing,
that I now call your attention further upon this
subject.   After the prisoner was arrested, and on
his way to the justice's office, he has conversation
with the officers who accompanied him, in which he in
substance states the same thing, and again, on the next
morning after he is committed to the jail of the county,
in conversation with his partner, Palmer, he admits the
shooting.   The other bit of evidence bearing upon the
question is that in the examination of the deceased we
find the bullet had entered his body under the shoulder-
blade and had passed nearly through and lodged in the
muscles of the breast.   That this bullet, in its passage,
had passed through the lungs and had cut the aorta,
and was the cause of death.   Upon examination, as
further evidence, it was found that a bullet was lying on
the floor at the base of the chimney, where the chimney
passed up through the sleeping room, and there was a
mark some three or four feet from the door, as if the
bullet had struck against the chimney and then dropped
down.   Another bullet was found in the pillow upon the
bed.   The first question which becomes necessary for

you to consider is the time when the first shot was fired.
In the statement made by the prisoner to his partner,
the next morning, we are told that he entered the room
where Lynch and his wife were lying upon the bed. He
tells us he felt over them and tried to get pepper into his
eyes. The next morning, upon an examination by the
officers and the coroner, pepper was found scattered over
the bed-spread. Then he tells us Lynch got up and
came for him, as he expressed himself, and he commenced to shoot. Lynch ran and he continued firing
until he emptied his revolver. He followed him down
the stairs and got out the slung-shot and tried to hit him
with it. I do not pretend to give the precise words,
gentlemen. I am speaking from memory. You will
exercise your own recollections as to the precise words
used, but that is it in substance. Now, it is contended
on the part of the defense in the case, and becomes the
first question which it will be necessary for you to
determine in your own minds, that the shooting was
done in self-defense. In other words, at the time the
shooting took place, Lynch was engaged in rushing
upon him ; that he supposed or had reason to suppose,
as he claims, that he was in imminent danger of being
seized by Lynch ; and it was under such circumstances
that he fired. It is claimed, however, on the part of the
prosecution, that the firing must have taken place before
Lynch arose from the bed ; that in the faces of the
parties who were lying there appear particles of powder ;
that the bullet which was used had penetrated the pillow sham and pillow case, and was found inside one of
the pillows ; that the shooting must have commenced
before Lynch left the bed, and that it was only after he
was aroused by the firing of the first shot, while he was
still lying there, that he jumped from the bed and ran
down stairs. Now, something has been said in reference to the number of shots that were fired. We have
had produced in court the revolver. We find, when the
revolver was taken from him, in one of the chambers

there was an unexploded cartridge ; that there were in the chambers four of the shells which had been exploded. Only three bullets have been found in the house. If there was a fourth one fired, they have not as yet been able to find it. The only evidence bearing upon the subject is the testimony of Swanson and the condition of those exploded shells found in the pistol of the defendant. Now, I do not regard it of great consequence whether there were three or four shots fired. We are unable to determine in the case which one of the shots it was which took effect upon the body of the deceased. The fact that it entered the body and passed through horizontally would seem to indicate that it was fired when the individual stood on a level with the person who fired the shot. It is claimed by the defense that the shot must have been fired before deceased commenced going down stairs, and the fact that there was blood found upon the stairs would seem to give force to the theory ; but that it was fired, and that it took effect, there is no contest. It is conceded by the defense : so that, gentlemen, the question which you are to determine is whether or not it was fired in self-defense, under such circumstances that the defendant had a right to suppose he was about to be assaulted and about to receive great bodily injury, and that there was imminent danger of his so receiving such injury, and he fired the shot to protect himself. If he did, gentlemen, then, of course, it would be your duty to render a verdict of not guilty. If he did not, if he commenced firing while the parties were lying upon the bed, and Lynch, after the shot was fired, jumped from the bed and ran down stairs, and while doing so the other shot followed, then, gentlemen, in case you should come to a conclusion from all the circumstances in the case that the prisoner was in the heat of passion and using a revolver, a revolver being a dangerous weapon within the meaning of the statute, then it would be your duty to convict him, unless you should become satisfied that another defense

21

has been established. Insanity, gentlemen, was stated by Dr. Mann to be a disease of the body affecting the mind. Dr. Andrews describes it as being a disease of the mind. The provisions of the statute upon these questions are as follows :

"A person is not excused from criminal liability as an idiot, imbecile, lunatic or insane person, or of unsound mind, except upon proof that, at the time of committing the alleged criminal act, he was laboring under such defect of reason as either not to know the nature and quality of the act he was doing, or not to know the act was wrong."

This, gentlemen, is the statute. It is in substance as has been previously described by our Court of last resort. By that Court it was defined to be "one laboring under such defect of reason from disease of the mind as not to know the nature and quality of the act he was doing, and if he did know it, that he did not know that he was doing wrong." The test of responsibility is the capacity of the person to distinguish between right and wrong at the time of, and in respect to, the act complained of. He must have sufficent reason or understanding, and have an intent to do wrong, and power to distinguish between right and wrong. The question therefore is : "Was the prisoner, at the time of shooting deceased, in such a state of mind as to know the deed was unlawful and morally wrong ?" If he was, then he is responsible. If he was not, then he is not responsible. This, gentlemen, is the law bearing upon the question of insanity.

Now, gentlemen, considerable discussion has taken place in reference to the evidence bearing upon the question of insanity on the part of the defense. It is claimed, in the first place, that insanity existed in the family of defendant, that was hereditary, and you will call to mind the evidence bearing on that subject. It is claimed that the father of the defendant has also suffered from epileptic seizures, or something that was of that form, and from which, it is claimed, the disease of insanity is

traced down through different periods, and that it exists
as hereditary in his family. It is for you to determine,
gentlemen, what force and effect should be given to the
evidence, and as to whether the claim of hereditary
insanity is well founded. It is one of the steps relied
upon by the defendant in the case. Again, it is claimed,
on the part of the defendant, epilepsy existed in the
defendant himself; that he had been subject to these
seizures in the early morning and while he was
yet sleeping. Upon this branch of the case the
testimony of his sister has been taken, and that
this form of epileptic seizures indicated a disease
of the mind. On the part of the people, it is
claimed, that these seizures which are described by the
sister were nothing more than ordinary nightmares, and
were not epileptiform. It is for you to determine, gentlemen, what the facts are in this regard, and as to the
force thati s to be given them, in case you find he has
suffered from epileptic seizures.

Again, gentlemen, it appears, from the evidence in the
case, that about the 17th of August, and after he had
returned from a business trip, his partner gave him some
information bearing upon the character of his wife, and
from that time to the 31st of October, they had frequent
conversations upon the subject, in which he was told that
his wife was not true to him, that she had been guilty of
intercourse with other men. It also appears from the
evidence that, during this time, he labored under great
sorrow, depression of spirits and mind, at times was seen
to shed tears or have red or swollen eyes. It is claimed,
on the part of the defendant, that in conversation with
some of the neighbors, there were some expressions made
in which he appeared to be absent-minded; appeared to
be in deep thought. Some of them characterized his
actions as irrational or unnatural. And again, it is
claimed that, down at the National Hotel, during the
two days that he remained there waiting for Lynch
to come up from Utica to visit his wife, his conduct

and appearance were not those of a rational, sane man; while on the part of the people, it is claimed that the condition of his mind, as described by various witnesses, and who spoke in reference to his conduct and action prior to that time, that they were only the natural actions of a man who loved his family, loved his wife, and had learned of her infidelity to him; that the depression of spirits, weeping, etc., that his absent-mindedness and other conduct, that has been described by the witnesses, were but the natural results of such a condition. It is for you to determine, gentlemen, what is in this, and as to the effect that should be given it. Again, after the shooting took place, and after he was incarcerated in the jail of the county, it is claimed then that he was melancholy, quiet, nervous, restless; that at times he broke out in fits of anger, and soon that he had a quarrel with his partner, Mr. Palmer, and from that, it is claimed, there was evidence of his laboring under disease of the mind. It is claimed also, from the manner in which he provided himself with weapons and other implements to go to his house and meet this man, that there was evidence of insanity. On the part of the prosecution, it is claimed, after he was taken to jail, that he was committed on the charge of murder, and that melancholia, the condition the man was found in from day to day, was but the natural result of a man who was resting under so grave a charge, and had but recently found out his wife's infidelity, and had slain her paramour. It is for you to take all this evidence together, before and after, and in your own way determine whether or not, at the time of the shooting, he was possessed of mental ability and of power to know that the act was wrong and unlawful. If he did know it, then, gentlemen, the defense is not established. If, however, you come to the conclusion, from the evidence, he was in that condition, did not know the shooting was wrong, did not know it was unlawful, then, gentlemen, he would be excusable. He could not be held responsible by our law. In addition to this, gentlemen, and as bearing upon this subject, we have the testimony of medical experts, gentlemen who are learned in the science of medicine and insanity, who have made diseases of the mind their study. They have been brought upon the stand and have been questioned in reference to their opinion in reference to assumed cases which have been embraced in questions which have been propounded to them. One of the physicians visited the individual in jail on the 15th of November, and made a personal examination of defendant, and he expresses the opinion that he was laboring under the disease of melancholia. Then, gentlemen, to aid you in determining the question of sanity or insanity, various symptoms of diseases have been given by these experts. Many of these are symptoms which may be looked for by ordinary men in examining a patient for the purpose of determining whether or not he was insane. Some of these symptoms you recognize as symptoms which would occur in other diseases than that of insanity. But, gentlemen, you may take the evidence as to actions, as to words uttered before and after the commission of the act : you are to take the circumstances into consideration as to the manner in which the act was committed, and then, gentlemen, you are to determine, according to your oaths and judgment, as to whether or not this man was laboring under disease of mind, and was he in that condition so as not to know this shooting was wrong and unlawful on his part? If so, gentlemen, then you must acquit him.

In conclusion, gentlemen, allow me te remind you that you are the chosen instruments which the law has provided to determine the guilt or innocence of the accused.

The people and the prisoner have the right that you should determine this question according to the law and evidence. Your Creator has endowed you with reason, judgment and discretion. You are to make use of your best abilities in the consideration of the evidence.

Maturely deliberate and conscientiously determine, to the end, that, in your verdict, justice may prevail.

At the conclusion of Judge Haight's charge, at 4.45 P. M. (just one hour from its opening), Judge Sutton, for the defense, made numerous requests of the court for charges on specific points, all but one or two of which were granted, and nearly all of which favored the defense.

Judge Bangs, for the people, asked the court to charge that insanity alone is not a good defense to crime unless it reaches a stage in which the accused cannot distinguish between wright and wrong.

The Court: I do so charge.

Judge Sutton: We object. (Exception granted.)

At 5.20 P. M. the court ordered Officers Tilley and Reynolds, who have had charge of the jury, to be sworn, and asked if the jury preferred to occupy its room in the St. James Hotel or in the court ante-rooms.

The jurors and officers preferred the hotel.

The court ordered the jury to the hotel and supper provided, if necessary.

Judge Haight then ordered Mr. Rowell to remain in the custody of Sheriff Southworth until a verdict was rendered, and he and his counsel cheerfully acceded to the order.

On leaving the court-room with Deputies C. H. Reynolds, of Elba, and James F. Tilley, of Oakfield, the jury went to the large sitting-room in the St. James that had been provided by Col. Collins. A proposition to take a preliminary ballot was rejected by the jury, which decided to go to supper first. Their table was adjoining that occupied by the representatives of the "Observer," who saw in the expression of relief upon their countenances, the heartiness with which they enjoyed their supper, and their easy method in conversation indicated the nature of the verdict. At 6.35 they concluded their supper and returned to their room.

Elijah Town, of the town of Alabama, the first man chosen on the jury, a clear-headed, reputable and prominent man in his section, was chosen foreman.

A proposition was made at once to ballot upon the question of "guilty" or "not guilty." There being no ballots handy there was a division of the house.

All of the twelve men took the not guilty side.

Immediately there was a proposition to divide the house on the questions, "acquittal on the ground of self-defense," or "acquittal on the ground of insanity."

All of the twelve men passed to the "self-defense" side of the house.

The jury then instructed Foreman Town to give a verdict of "Not guilty."

The actual time occupied in these proceedings was just one minute and a half.

Then the jurors adopted a resolution of thanks to Officers Reynolds and Tilley for their faithful and courteous attentions during the trial, and prepared to return to court at 6.55 P. M.

Information was given Judge Haight that the jury was ready to report, and the court was ready for business at 7. At 7.05 the jury came into court. The names were called, but Rowell did not respond — the first time that he had been late since the trial opened.

Sheriff Southworth's brother was sent after the sheriff, and he and Rowell entered the court room at 7.20 P. M. Rowell appeared the most unconcerned man in the room. He told the "Observer's" representative that he was ready to return so as to be on time, but did not think it would be good taste for him to hurry the sheriff.

THE VERDICT.

At 7.25 P. M. the county clerk said : Gentlemen of the jury ; have you agreed upon a verdict?

Foreman Town (rising): We have.

Clerk : How say you. Do you find Edward Newton Rowell, the prisoner at the bar, guilty or not guilty?

Foreman Town : We find him not guilty.

Hardly had the word "not" issued from the lips of the foreman when a double round and whirlwind of applause and cheers came from the audience, which was composed almost wholly of people of Batavia.

Judge Haight struck his gavel on the bench, but it could not be heard. He waited until the last round of applause was ended, and then, striking his gavel once, said: "The audience will please come to order." Silence reigned at once.

In the meantime, Rowell's father, brother, uncle, Charles Steele, the faithful friend from Brooklyn, Judge Sutton, Counselor Watson and the reporters had all grasped the hand of the modest and quiet little man, who seemed the least concerned of any one.

The clerk asked the usual question of the jury as a whole: "So say all of you ?" and received the responsive " Yes," and affirmative nods of the heads.

Judge Haight : In finding your verdict did you pass upon the question of insanity ?

Foreman Town : The jury found that the defendant acted in self-defense.

Another outburst of applause followed this announcement.

Judge Haight : Under the circumstances, then, the court will not detain the accused any longer. Mr. Rowell, you are at liberty.

Rowell : Thank you.

The last question of the court was to determine whether, under the finding of the jury, it would be necessary to appoint an expert commission to decide upon Rowell's mental condition, and if found insane, he would have to be committed to an asylum, to remain until adjudged sane or safe to be at large.

A great deal of the success attending the defense of Rowell was due to the indefatigable efforts of Judge Wm.

B. Sutton, of Utica. N. Y., who conducted the case as
senior counsel, and who showed a familiarity with
mental diseases and their symptoms rarely met with
outside of the medical profession. His management of
the case was throughout a masterly legal effort, which
won instant recognition from all appreciative listeners.

The jury in this case were not willing to bring in a
verdict passing on the question of insanity, as they feared
Rowell would in that event be remanded to an asylum
for the insane.

CHAPTER XV.

THE PSYCHOLOGICAL ASPECT OF THE CASE OF LUCILLE YSEULT DUDLEY.

THE following is the history of the case of Mrs. Lucille Yseult Dudley, who recently shot O'Donovan Rossa with the intention of killing him, imagining that he was the instigator of the dynamite outrages, which startled London. We examined the prisoner at the request of counsel for defense, and gave testimony respecting epilepsy, as producing irresponsibility.

Examined Lucille Yseult Dudley in the Tombs, New York, April 24th, 1885. She was a woman suffering from the disease of epilepsy, and also from congenital moral insanity, or, emotional insanity proper. She had a tendency to delusive or insane opinion, and to the creation of morbid or fantastical projects. She was a woman who had been in a more or less of a morbid mental state throughout her life, and was probably insane at the time she shot O'Donovan Rossa. Physical condition good. Memory good. Eyes have an insane expression.

The defense of insanity in this case is made out by most clear and convincing proof, as follows, viz. :

We have a case exhibiting an exceptionally quick intelligence and decided power of discrimination, together with a chronic condition of insanity before the crime.

Her condition is the result of heredity, being transmitted from her maternal grandmother, who was a case of suicidal mania. When married she suffered from the strong moral shock of discovering that her supposed husband was really the husband of another woman, a fact well calculated to induce insanity in any susceptible woman of naturally poor or weak mental balance. She immediately separated herself from this man. Her child, whom she loved, dies at the age of three years. Alone, forsaken, deserted, with a family predisposition to madness, suicidal mania now manifests itself and she attempts self-destruction. Is now admitted to Dr. Williams' asylum, in England, where she remains seven months, and is then discharged.

She suffers meanwhile from epilepsy, which is sufficient alone to produce complete irresponsibility. [See evidence given in Nelly Vanderhoof case by Dr. Mann, before Judge Van Brunt, with the latter's charge to the jury.]

Her mental powers had become impaired as the result of epilepsy, and she had the irritable condition of the nervous system produced by epilepsy, and epilepsy was the phase of mental disturbance that prompted the criminal act of shooting Rossa.

During her past life she had been many times under the dominion of that blind fury so frequently exhibited by epileptics, both before, after and between the fits. Her mind was generally so impaired that she was seemingly incapable of controlling the feeblest impulses of passion. She was laboring under a disease which almost inevitably impairs the mind.

She had congenitally feeble moral powers, a moral insensibility and necessarily a proportionate irresponsibility. This congenital deficiency is the result, probably, of imperfect nutrition of the textures of the brain, occurring, perhaps, even during fœtal life. Perfect sanity has never, in our opinion, existed, as we do not consider that her brain and nervous system have ever been in a condition that the mental functions of *feeling and knowing, emotion and willing,* have ever been performed in their regular and usual manner.

"Insanity means a state in which one or more of the above named functions is performed in an abnormal manner, or not performed at all, by reason of some disease of the brain or nervous system." [See Stephens' Criminal Law, 1883, vol. II, p. 130.]

She is a case of moral insanity or emotional or reasoning madness and epilepsy combined. Such a case is reported at p. 244, Bucknill & Tuke's Psychological Medicine, 1879. She as much requires guidance, restraint and treatment as the most furious maniac. She is thoroughly unconscious of ever having done anything wrong. Like all other subjects of emotional insanity proper, she cannot control her feelings. There is no delusion. Without epilepsy she would have a condition of the affective power of the mind which is so deficient as to lessen responsibility. Without emotional insanity, she would still remain irresponsible from epilepsy. She has, in her life, exhibited emotional irregularities rather than delusion or hallucination, but has none the less labored under cerebral disease. While she talks quite rationally, she shows by her acts and conduct that she is mentally deranged. She has the condition of emotional insanity, in which the mental disorder is of a sudden and transitory character. The duration of the morbid state is short and its cessation sudden. Her outbursts of maniacal fury and destructive and homicidal impulses are of this nature. She carries within her the active organic influence of a morbid nature, which, although not extremely noticeable when disturbed and disordered at the moment of action, turn the scale towards crime.

We come finally to the most important part of this case, viz.: the exculpatory effects of the disease of epilepsy and its medico-legal relations. While there is existent among the laity a disposition to dispute the existence of emotional insanity proper, or moral insanity, there is no well educated physician in the country who does not know that the disease of epilepsy produces a modified responsibility in all the subjects of said disease.

"The subtle influence of epilepsy is remarkably manifested in the change which takes place in the moral character, either permanently, or during brief periods. In a large number of cases, the actual or comparative sanity of the patient for considerable intervals of time, the freedom from irascibility, passion or violence, when removed from circumstances calculated to irritate, render it difficult to place such persons under restraint until an act has been committed which necessitates sequestration." [See Bucknill & Tuke's Manual of Psychological Medicine, p. 336.]

Very often the character of the mental disturbance, the paroxysmal gust of passion, the blind fury without an adequate cause, indicate the presence of epileptic insanity, and take the place of epileptic fits. Masked epilepsy is indicated by eccentric acts or a sudden paroxysm of violence, without a distinct epileptic seizure. Unmistakable epileptic fits occur at one period of a patient's life, while at another, maniacal symptoms take their place. When mental symptoms appear to take the place of a fit, there is transitory epileptic paroxysm. All acts after epileptic fits from vagaries to homicidal actions are automatic, and the patient is irresponsible.

Elaborate and complex actions may be performed while a patient is unconscious. In different cases there are different degrees of recollection, *as in other forms of insanity there may be a motive mixed up with an insane condition.*

There may be motive and calculation in some cases, which, in some *rare* cases, control the misdeeds of epileptics. *Echeverria* says, "for an alienist it is certain that the victim of a disease which takes away from him all control over himself, even when he remains capable of distinguishing between good and evil, cannot be held responsible for acts which he accomplishes without will, and in an automatic and therefore unconscious manner."

"There is no epilepsy without unconsciousness. Epileptic seizures vary in severity from a simple vertigo, scarcely discernable by others, to the most violent convulsive fit, lasting from minutes to some hours." [Ray, Medical Jurisprudence of Insanity, 5th ed., p. 474.]

"Anger, fright, or any strong moral emotion is very liable to produce a paroxysm. Epilepsy tends almost invariably to destroy the natural soundness of mind. A direct, though temporary effect of the epileptic fit is to leave the mind in a morbidly iritable condition, in which the slightest provocation will derange it entirely." [Ray, Med. Jurisprudence, p. 475.]

This is precisely the state that Lucelle Yseult Dudley was in when she shot Rossa. She had within a few days had such an attack, and the provocation was the news which arrived from London of the dynamite outrage, of which she imagined Rossa to be the direct instigator. *Her criminal act was the result of the morbid irritability which succeeded her epileptic paroxysm.*

Previous to this time, *about three weeks*, her friends where she boarded had expressed views as to her irresponsibility. The crime was the result of an abnormal condition of the nervous system.

"In epileptics it is not uncommon to observe attacks of mania which are characterized by a high degree of blind fury and ferocity." [Greisinger, Mental Pathology and Therapeutics, 2nd ed., London, 1868, p. 289.]

"During the attack the patient is unconscious, so that his acts, whatever may be their nature, cannot make him liable to legal punishment. The passionate impulse to kill in masked epilepsy is substituted for ordinary epileptic convulsions. Instead of a convulsion of muscles, the patient is seized with a convulsion of ideas." [Responsibility in Mental Diseases, 4th ed., 1881, pp. 66-70.]

"An epileptic convulsion may not occur, but may be represented by sadness, dejection, by sullenness, by ebullitions of rage and ferocity, a *mania transitoria*, signalized by suicide, homicide, and every modification of blind and destructive impulse. The awakening from epileptic stupor may often resolve itself into an outburst of mental derangement, manifested by extreme vehemence, violence and destructiveness." [See Mann's Manual of Psychological Medicine and Allied Nervous Diseases, 1883, p. 306.]

A crime resulting from epileptic psychical phenomena may be accomplished with comparative deliberation, and, as we have before remarked, there may be a motive mixed up with an insane condition.

All epileptics are impressionable and excitable, and epileptic attacks are often replaced by irresistible homicidal tendency. A patient may recognize his impulses as illegal, but irresistible. In epilepsy dreamy mental states and imperative acts appear and disappear with great suddenness. *If the prisoner did premeditate the act and called upon Rossa with the intention of shooting him, that would not prove that she was not insane, or that she could control her insane desire ; on the contrary, it might be a still stronger proof of her insanity, that under the circumstances in which she was placed, she would do an act from the fearful consequences of which it was impossible for her to escape. Every day there are examples in insane asylums of insane persons committing crimes that they have premeditated.* Premeditation is no proof of a person's sanity.

I do not hesitate to declare Lucille Yseult Dudley a woman of unsound mental organization. If she shot Rossa in the manner in which she is said to have done, she shot him while laboring under an insane, epileptiform, uncontrollable impulse, for which she was not responsible, and at the time of the shooting she was not in a condition to realize the nature and quality of the act she was doing, or to know the act was wrong.

Finally, her volitions, impulses and acts have been determined by insanity.

She comes of a stock whose nervous constitution has been vitiated by mental disease.

She has been noticed to display mental infirmities and peculiarities, due both to hereditary transmission and to present mental derangement.

She has not the ability to control mental action, and she has not sufficient mental power to control the sudden impulses of her disordered mind, and she acts under the blind influence of evil impulses which she can neither regulate or control.

Her act of shooting Rossa was accomplished without an adequate incentive or motive, or in other words, a sane person would not have considered Rossa's conduct as excusing them for killing him.

She has exhibited depression and excitement; moody difficult temper; a habit of unreasonably disregarding ordinary ways, customs and observances; an habitual extravagance of thought and feeling; an inability to appreciate nice moral distinctions, and, finally, she gives way, by reason of epilepsy, to uncontrollable gusts of passion and blind fury. These mental defects are, taken together, unmistakable signs of insanity.

In Conclusion. Homicide, or assault with intent to kill, is not criminal, in our opinion, if the person by whom it is committed is, at the time when he commits it, prevented by any disease affecting his mind from controlling his own conduct, or, as Bucknill prefers to put it, if Lucille Yseult Dudley, at the time of the shooting, was suffering from " *incapaciting weakness or derangement of mind produced by disease,*" then she was insane, and we consider that she was so suffering.

Hypothetical Questions of Defense. "Take the case of an individual descending from enfeebled parentage; derived from a race in which insanity was hereditary; the victim from childhood of frequent epileptic seizures, occurring all through infancy and childhood, and later at frequent intervals; who, after months of intense agony, emotional disturbance and sleepless nights,

struggling with alternate hope and fear concerning
the probability of her supposed husband being really
the husband of another woman : when finding such to
be the case, separates from him : whose child dies ; who
then in despair attempts suicide, and is taken to an
insane asylum ; whose judgment and mental powers have
become impaired ; who has become, as the result of disease of the brain, epileptic, irascible, passionate and violent, and who was seen frequently to manifest paroxyoms,
gusts of passion and blind fury without an adequate
cause ; who has been heard to express suicidal and homicidal ideas ; acting in an unnatural and irrational manner ; who becomes much excited at hearing of outrages
to her country, and takes a revolver, and calling upon the
man she imagines to be the instigator of said outrages,
fires a shot at him ; who immediately after the shooting,
was calm, made no attempt to escape or to deny the
shooting; who was careless, unconcerned and unmoved
in the midst of intense excitement; and who, during
some days prior to the shooting, had taken but little
nourishment, and was sleepless, and who had suffered
much previous to the shooting from epileptic attacks :
was that individual at the time of the shooting, in your
judgment, laboring under such a defect of reason as not
to know the nature and quality of the act she was doing,
and not to know that the act was wrong ?"

Answer : "She was."

We are credibly informed that Mrs. Dudley attempted
suicide during the month of June while in Jefferson
Market prison, and she has freely aired homicidal
ideas, even threatened to kill certain public officers, provided she was "hounded" on her trial. She has also
expressed the hope that she might be convicted and
sentenced to the penitentiary, "as it would be better for
the cause."

The trial commenced in the Court of General Sessions,
June 29, 1885. The prosecution put Rossa on the stand
to testify, and then commenced a scene rarely seen in

22

any court-room. Mrs. Dudley became very much excited and hurled such an avalanche of invectives and torrent of abuse at the witness, as to upset all order and decorum in the court-room. Two officers were stationed on either side of her, but could not keep her quiet. "Liar," "scroundrel," "assassin," "coward," etc., were frequently interjected by her during his testimony, and she kept up a running fire of scathing remarks during the whole time he staid on the witness stand.

Testimony was introduced showing the prisoner's insanity and incarceration in Hayward's Heath Asylum, in England. The prisoner then insisted on taking the stand and making a statement, which she did. The writer was then put on the witness stand as the expert for the defense, and testified to the past and present insanity of the prisoner. The jury, after being addressed by the judge, retired, and in five minutes brought in a verdict of not guilty on the ground of insanity.

These trials all tend to show the duty of the prompt and early seclusion of the insane away from home. There are too many insane people who are outside of asylums. To leave an insane man wholly to himself is, as these trials show, a very dangerous course for society to adopt. There should be provision in every State whereby a person standing on the border line of insanity can receive the benefits of asylum treatment without actual commitment as an insane patient. Patients are also removed too frequently from hospitals before cure, and then commit other overt acts. The superintendent of an asylum has no object in detaining a patient after he is cured, and the public and judges and juries should accept as final the superintendent's opinion as to a cure, especially when backed by opinion of those skilled in psychiatry.

CHAPTER XVI.

Psychological Aspects of Three Cases of Infanticide Considered in their Relations to Forensic
Medicine.

These three very interesting cases, which the writer
studied very thoroughly, for the reason that he appeared
in each of the trials as the expert for the defense, were
conducted on different legal bases, and involved three
distinct points in law and medicine, aside from general
considerations on infanticide, which of course came out
in these trials. These cases are all of interest, involving,
as they do, questions likely to be asked of the medical
jurist or expert at any time in similar cases.

The first case is that of The People v. Nelly Vanderhoof, tried before Judge Van Brunt, at the New York
Oyer and Terminer, April 9th and 10th, 1885. The
defense was epilepsy. The second case, that of The
People v. Kate Harvey, in the same court. The defense
in this case was the insanity of seduced and deserted
women. The third case was that of The People (State of
Pennsylvania) v. Miss Le Bar, tried before Judge Howard
Reeder, at Easton, Pa., October, 1885. This case was
ably defended by Hon. W. H. Kirkpatrick, of Easton,
Pa., on the ground that the child was not born alive, and
that it had never breathed, as alleged by the prosecution.
It was claimed by the defense that; first, there were no
evidences of live birth prior to and independent of respiration ; and, second, that there were no evidences of live
birth subsequent to and deduced from respiration.

In two of these cases the verdict of the jury was for acquittal and for the defense. In the second case, that of Kate Harvey, unfortunately for the prisoner, I was not applied to until after the trial was over and the prisoner about to be sentenced. I then examined her and accompanied her counsel to court and pleaded a modified responsibility for her, with the effect of obtaining for her a very light sentence. The defense might have won their case had they introduced expert testimony on the trial ; but as the woman was absolutely friendless and very poor, the case was probably hurried over.

CASE I. *The psychological aspects of the trial of Nelly Vanderhoof, held in New York, April 9th and 10th, 1885, for the murder of her newly-born babe. Medico-Legal Relations of Epilepsy.*

During the latter part of the month of November, 1884, the defendant, Nelly Vanderhoof, aged twenty-two years, gave birth to a child. She had suffered from epilepsy from birth ; the family were saturated with this disease : she was a young, unmarried woman, suffering from the strong moral shock of seduction and desertion, and the irritable conditions of the nervous system produced by epilepsy ; she also had, when we first saw her, a considerable degree of uterine derangement. When the writer examined her at the Tombs, she had apparently no realizing sense of the enormity of her crime, and the mental powers had become very obviously impaired as the result of epilepsy. After investigating her mental condition, the writer reported to her counsel that she was, in his opinion, irresponsible, and that epilepsy was the phase of mental disturbance that prompted the criminal act. During her past life she had been many times under the dominion of that blind fury so frequently exhibited by epileptics, immediately before or after a fit. Her mind was usually so impaired that she was seemingly incapable of controlling the feeblest impulses of passion. She was laboring under a disease which almost always impairs the mind ; she has a sister demented as the result of the same disease, a resident of one of the New York institutions for the insane. Her father is a case of dipsomania ; her mother has twice attempted suicide. Such was the prisoner's mental condition and her family history. Her trial took place before Judge Van Brunt, in the New York Oyer and Terminer, April 9th and 10th, 1885. The people were represented by Assistant District Attorney Fellows, who, in trying the case, deserves great credit for his enlightened and humane views respecting the exculpatory effect of the disease of epilepsy. In this trial the honorable district attorney not only did not, by his professional act, try to deprive the expert expression of opinion of its proper weight with the jury, but showed every courtesy, and went so far in his efforts to elicit the whole truth and the matured convictions of the expert as to suggest to the counsel for the defense that he would not object to Counselor Bailey asking Dr. Mann *for his opinion, founded on all the evidence given at the trial, supposing it to be true.* This was a step in the right direction, and showed a liberal and enlightened spirit on the part of the eloquent and able assistant district attorney. During the whole trial this gentleman proved himself as humane and progressive in his ideas of the psychical states in epilepsy as has Sir James Fitzjames Stephens, of England, in his lately published "Criminal Law," 1883, vol. II, p. 141. On the trial there was no attempt on the part of the counsel for the defense to break down the testimony which proved the killing of the newly-born babe. For the defense the following testimony was elicited from the father and mother : Two aunts and an uncle were epileptics ; that a sister Emma, who died at the age of thirty, was an epileptic ; that there is a daughter living, and in confinement on Randall's Island, who is imbecile and epileptic from birth. Both father and mother testified that the prisoner, Nelly, had epileptic fits during her infancy and childhood ; would cry out and lose consciousness, and fall down.

Later in life had these attacks with every menstrual period; had always had impulses to violence, and would threaten people and strike them; would have attacks of blind, irresistible fury, without any adequate external exciting cause; would threaten her mother; would become violent, incoherent and destructive at times, between the fits; were afraid of her at these times; during her pregnancy, was wild and incoherent at times, and violent.

Adelaide Sullivan, of 2290 First avenue, was sworn, and testified that she knew Nelly Vanderhoof three years ago. She acted very peculiarly, sometimes acted very nicely, and then would become very angry and violent when no adequate cause existed for such action. "Did not become intimate with her; as I thought her queer." The witness made up her mind that Nelly was crazy; had heard her talk fast and incoherently; her eye was wild and rolling at such times. She acted without apparent motive; had heard Nelly talk about cutting people's throats and express homicidal ideas.

Mrs. L. Kesslar, of 2290 First avenue, was sworn, and testified that the prisoner, Nelly, lived with her three months in 1882. Never thought she was in her right mind; called her "Crazy Nell." Without adequate cause she would get into a terrible temper and talk incoherently and make wild gestures; have heard her make homicidal threats repeatedly; would complain of sleeplessness and headache; she would always lose her mental balance upon the slightest occasion. Both of these witnesses testified to having witnessed unusual and irrational conduct at times, and inchoherence of speech on the part of the prisoner. The writer was then sworn as the expert for the defense, and testified substantially to the following facts: Examined N. V. in jail this A. M., and thrice previously; head hot, eyes suffused, face flushed, pulse full, hard and ninety, skin dry and harsh, tongue has been bitten frequently; examined for uterine trouble; found retroversion, with evidences of previous pelvic cellulitis; complains of pain

on top of head, which had lasted for some time; pain
and dragging in lower part of the back ; pain in the
thighs, and leucorrhœa ; think she has had sufficient
uterine derangement to produce some cerebral irritation :
consider her an epileptic with transitory homicidal
impulses in the past and probably in the future :
epilepsy hereditary on father's side ; mother a neurotic
woman, with melancholia associated with suicidal ideas :
that epilepsy is a disorder of the nervous centers, the
phenomena of which morbid state consist in seizures
generally sudden in their invasion ; preceded as a
rule by well marked paroxysms, characterized by loss of
consciousness (coming on suddenly), and attended by
peculiar involuntary muscular movements, which are
highly spasmodic and convulsive in nature : that there
may be in this disease, loss of consciousness without
evident spasm ; that there may be loss of consciousness
with local spasm only ; that the cause of epilepsy is pre-
eminently hereditary taint; there is generally a family
taint present; that epileptiform attacks may be partial
in nature and may not reach convulsive activity except
as far as the mind is concerned : that there is an epileptiform state which manifests itself chiefly by irritability,
suspicion, moroseness and peevishness of character, with
periodical attacks of maniacal fury ; that all kinds of
doings after epileptic fits, from slight vagaries to homicidal actions, have one common character — they are
automatic, they are done unconsciously, and the patient
is irresponsible : there is mental automatism : that elaborate and highly compound actions may be performed
when a patient is unconscious. With respect to epilepsy, it is absurd, as Echeverria has truly stated, to
suppose that motive and calculation imply necessarily
free will or soundness of mind : that great stress should
be laid upon nocturnal attacks and vertigo; that the
latter is more injurious to the integrity of the brain than
any other symptom ; that very frequently the presence
of epileptic insanity is indicated, not by epileptic fits,

but by the character of the mental disturbance, the
paroxysmal gust of passion, the blind fury without an
adequate cause; that mere epileptic vertigo or *petit mal*
is quite as dangerous to the integrity of the brain as the
grand mal, and even more so ; that there is a masked
epilepsy marked by eccentric acts or a sudden paroxysm
of violence without a distinct epileptic seizure ; that
there are in epilepsy, from time to time, attacks of
mental excitement accompanied by homicidal impulses,
which appear to take the place of the ordinary convulsive attacks ; that when mental symptoms appear to
take the place of a fit, there is, as Hughlings Jackson
has authoritatively stated, a transitory epileptic paroxysm ; that Delasiauve, Trousseau, Falret, Morel, Echeverria, Baillarger, Castlenau, Tardieu, Dagonet, Browne,
Jackson, Ray, and Bucknill and Tuke all agree as to the
irresponsibility of epileptics.

HYPOTHETICAL QUESTION FOR THE DEFENSE.

Take the case of an individual descended from enfeebled parentage — derived from a race in which insanity
and epilepsy were hereditary, the victim from birth of
diurnal and frequent nocturnal epileptiform seizures,
occurring all through infancy and childhood, and later
accompanying each menstrual period, who, after months
of intense mental agony, emotional disturbance and
sleepless nights, suffering meanwhile with a disease
peculiar to women, struggling with alternate hope and
fear concerning the probability of a promise of marriage
which had been made, being kept ; suffering from constant headache ; whose judgment and memory have
become impaired ; who has become, as the result of disease of the brain (epilepsy), irascible, passionate and
violent ; who was seen to present a wild and staring
gaze, and who was seen frequently to manifest paroxysmal gusts of passion and blind fury without an adequate cause ; who has been heard frequently to express
homicidal and suicidal ideas ; failing to give ordinary
recognition to conversation, when met at various times ;

acting in an unnatural and irrational manner, incoherent at times; who passes through the throes of pregnancy and parturition; who, after going up and down stairs repeatedly in an automatic manner, finally sees a knife, and kills her new-born babe; who, immediately after the killing, was calm and absent-minded, made no attempt to escape or to deny the killing; who was careless, unconcerned and unmoved in the midst of intense excitement; and who, during some days prior to the killing, had taken but little nourishment, and was sleepless, and who had suffered much previous to the killing from epileptic vertigo; was that individual, at the time of the killing, in your judgment, laboring under such a defect of reason as not to know the nature and quality of the act she was doing, and not to know that the act was wrong?

Answer: She was.

The trial judge charged the jury that they had heard all the evidence in the case tending towards the exculpatory effect of epilepsy on the mind, and, that if they believed it to be true, it was manifestly their duty to return a verdict of "Not guilty," on the ground of insanity. This the jury did after five minutes' deliberation, without leaving their seats.

CASE II. *The People v. Kate Harvey, for the murder of her newly-born babe. The court acknowledges a modified responsibility as attaching to women distracted by conflicting feelings, guilty of killing their newly-born offspring, when they have been seduced and deserted, and in consideration of this fact, mitigates the intended severity of his sentence.*

In this case, we have a young girl of nineteen years of age, without the safeguards of home and friends, seduced and deserted by the wretch who had promised to marry her, and who, torn by the distracting feelings incident to her shame and to parturition, undoubtedly drowned, in the bath-tub, her newly-born babe. Such women are hardly responsible for their acts we think, and had we been called upon the case early enough, we should have hoped to have been instrumental in securing an acquittal. As has been stated, however, the counsel made the fatal error of not introducing any expert testimony on the trial, and it was only when the jury had found a verdict of "Guilty," and the girl was to come up for sentence that I was applied to. I then proceeded to the Tombs and examined the prisoner, with the following result: She was nineteen years old, and of neurotic parentage ; her mother had been a life-long sufferer from cranial neuralgia ; consumption and rheumatism were plenty in the family. In my work on "Mental and Nervous Diseases," I have spoken of the correlation of morbific force in diseases, and that we often find phthisis and insanity running in and out in the same family.

The prisoner first menstruated at the age of twelve years ; always had experienced great pain at the top of the head at this time and much uterine pain ; there were always pains in the back and limbs, of considerable severity ; was always sleepless at these times.

About a month previous to the trial she acknowledged killing her baby ; said that she did not intend to kill it ; that her confinement came unexpectedly, and that she had a great deal of confusion of thought, and "that her brain was in a whirl ;" said that she was taken with labor pains in the bath-tub, and could not get out ; said that she tried to get out, but that the child was dead before she fully realized what had happened. Then she was afraid, and hid the child in an ash barrel. She felt sorry about it, and would rather the child had lived than not ; said that her statement was taken in the hospital, and that she had so much pain and mental distress that she was not conscious of what passed at that time. It seemed now as if it were all a dream. She lived in New York, and was seduced under promise of marriage. She was sleepless and had night sweats when we examined her ; appetite very poor ; the head was hot; said that she suffered very much mentally during all her pregnancy; was educated in the Protestant faith, and knew that murder was wrong, but was evidently in such a mental state at the time that, even supposing she had intended to kill her child, which she claimed she did not, she would have not been in a mental state where a calm consideration of right and wrong could have been possible.

We accompanied the prisoner's counsel to the court, and although she was about to be sentenced, as the jury had rendered a verdict of guilty, were, by the courtesy of Judge Barrett, a man of habitual fairness, allowed to make a statement in the prisoner's behalf. We claimed a greatly modified responsibility in these cases, and claimed the existence of the insanity of seduced and deserted women, as in many cases not only existing, but prompting to acts of suicide and infanticide. Judge Barrett then said, that while such a statement should have been made during the trial and not after it, and while, after the verdict of guilty, he must do his duty and show the public that infantile human life must be protected and infanticide severely dealt with, yet, in view of all the facts of the case, he would pronounce a very different sentence from that he had anticipated, and accordingly did so, giving a very light sentence. The trial judge in this case evidently felt that there was a degree of irresponsibility here that deserves the protection of the court, and had expert evidence been introduced at the proper time, no jury would ever have convicted the prisoner.

CASE III. *The People (State of Pennsylvania) vs. Miss C. Le Bar. This case was tried in Oct.*, 1885, *at Easton, Northampton Co., Pa., before Judge Howard J. Reeder.*

The prosecution in this case proved that a child had been born and that its birth had been concealed. These facts were not denied by the defense. The prosecution, moreover, claimed that the prisoner had murdered her newly-born babe. This the defense denied, claiming that the child was not born alive.

The questions brought up were, the degree of maturity of the child; the question as to the child being born alive; if alive, how long did it survive its birth? how long it had been dead when found, and the cause of its death.   The great question with the defense was : Was the child born alive? and they claimed not.

The child was found in a vault behind the house. There were no marks on the body of the child, which could have been inflicted while the blood was still circulating, and the prosecution did not claim that death had been produced in any other way than by throwing the child in the vault to die.   The prosecution relied as a test of live birth upon the·hydrostatic test, or the buoyancy of the lungs when placed in water.

The prosecution also asserted extra-uterine life from the fact that the umbilical cord was found to be shrunk and mummified.

The writer claimed for the defense, that the latter fact was not of the slightest value as a proof of extra-uterine life.   That it happens with portions of the cord cut off and exposed; that it was not a vital process.   Respecting the allegation of the prosecution of the fact of buoyancy of parts of the lungs as conclusive proof of live birth, we objected, for the defense, that the buoyancy of the lungs may be due, not to respiration, but to emphysema, to putrefaction or to inflation.   We claimed that many illegitimate children are born dead from natural causes; that protracted labor, premature birth, congenital want of power to breathe, loss of blood before or after birth, and compression of the cord, may cause death before, at, or soon after birth : that malformation or diseases of important organs lead to the same result, and that no proper examination was made by the prosecution to see if such a thing had happened; that the child might easily have died of neglect during its birth : that no woman is competent to attend to her child, alone and unattended, as this woman did, and that it had been proved that the prisoner was taken in labor, alone and unattended.

We asserted that the marks of respiration were wanting; that there was no testimony on the part of the prosecution to prove the appearance of developed air cells on the surface of the lung, which is characteristic, and which furnishes undeniable proof either of respiration or inflation; that these groups of developed air cells are bright vermilion-colored, and that it was the only lung test to which no serious objection can be offered. The witnesses for the prosecution testified that the true skin was more or less extensively discolored, and that the surface of the body was slippery. We testified that we should expect just such a state of things when a child had died in the womb; that these were marks of intra-uterine maceration. We showed that in imperfect respiration the changes in the size and shape of the chest (becoming after respiration larger and rounder) did not occur, and had no independent value as a test of respiration, as also the change in position of the diaphragm (becoming flattened and depressed instead of arched and high); that also altered positions of the lungs and altered consistence of the lungs, although corroborative of other evidence, had no independent value as tests; that respiration might take place before delivery and complete separation, and yet the child die during the delivery at or shortly after birth.

That the stomach in still-born children is lined with a glairy mucus, free from air-bubbles swallowed during the establishment of respiration; that later it may contain milk or farinaceous food, proving that the child was born alive and had lived long enough to be fed. The milk may be identified by the microscope and by Trommer's test, as used for detecting sugar in urine. This test gives characteristic results with the whey and curd of milk; that the large intestines in mature still-

born children are filled with meconium. Its complete
expulsion would afford a probability that a child had
survived its birth.

The question as to how long the child survived its
birth was not touched upon.

We testified that the changes in the cord of the child
were merely the common consequences of putrefaction,
and of not the slightest value as a proof of extra-uterine
life; that the best test of extra-uterine life, as regards
the cord, is the presence of the bright red ring surrounding the insertion of the umbilical cord, and that no such
appearance was testified to by the prosecution in this
case. We also testified, in conclusion, that the cause of
death might be due to the infant being immature and
feeble, that the infant might encounter obstacles to the
continuance of respiration, even if it had been alive,
or that in a similar event congenital disease might
shorten life; that the congenital disease in any given
case might have its seat in the heart, the lungs or the
brain.

The prosecution did not claim that death was due to
violence, so that there was no occasion to look, as we
sometimes have to do, for punctured wounds of the
fontanelles, orbit, heart or spinal marrow, dislocation of
the neck, extensive fractures of the bones of the head, or
for suffocation or strangulation, the signs of which are
apparent to a skilled observer in any given case. On the
other hand, the defense did not deny that the woman
had been recently delivered; so that this question and
the allied one as to whether the period of the woman's
delivery corresponds with the time at which the child is
supposed to have been born, were not, in this case, at all
for discussion. The question of puerperal insanity was
not raised, although really in this case the state of the
mother's mind was somewhat problematical. The jury
returned a verdict of "Not guilty" without leaving
their seats.

The person whose mental unsoundness is suspected should

1. Be examined as to his past history and that of his family. Insanity is generally hereditary, although not necessarily so.

2. The examination of the patient, if not showing hereditary insanity, may show other neuroses or nervous diseases, such as epilepsy, chorea, paralysis, hysteria, vertigo, etc., either in the suspected insane man or woman, or their family.

3. There may have been a change of habits or disposition, the result of injuries to the head, sunstroke, fevers or syphilis. The existence of disease of the lungs, heart or kidneys is also important.

4. Examine as to the mode or manner of the crime. Many sudden, motiveless, peculiarly atrocious murders are the work of epileptics. Did the criminal attempt to conceal the crime or himself?

The psysiognomy, attitudes and gestures, the words, manner of speaking and writings, the physical condition as to sensation, muscular power and organic functions, should all be carefully examined by the expert who examines a criminal for insanity; and if on a trial, the lawyer for the defense, on cross-examination, elicit the fact that the experts for the prosecution have neglected such inquiries, they should, and properly, claim that evidence based on an examination ignoring such careful procedure is untrustworthy and not entitled to weight with the jury. It is preposterous to see, as lately happened in New York, a man who had never personally examined a criminal, testify against him when on trial for his life. Such things should not be allowed. What is wanted is an expert opinion founded on a personal examination of the criminal in whose behalf the plea of insanity is raised, and also on the whole evidence of the trial, provided the expert has heard it all, as he should do.

The physician's examination should be very searching as to the prisoner's mental condition, and as to amatory ideas, religious ideas, ideas of property, ideas of an ambitious kind, and as to ideas on social subjects. The memory should be very carefully tested.

The lawyer who proposes to defend a criminal case, and who alleges insanity as a plea for defense, should learn, as thoroughly as possible, the whole life-history of the criminal. He should ascertain whether there has been a change in the habits or disposition. He should, through a physician, always have the most careful examination of a probable insane patient; peculiarities of residence or dress ; the appearances, demeanor and general conduct of a person ; the peculiarities of their bodily condition. The peculiarities of gesture and expression of the countenance will generally, to an expert, show clearly either sanity or insanity, if not on one visit, after a few careful examinations.

We can see no objection, and can see many advantages, that would result both to society and to the insane, if, in criminal cases where insanity is to be the defense, the disease still existing or being claimed to exist, the court should, by legislative enactment, order the prisoner to be placed in the State Hospital for the Insane, that he might be under medical observation.

CHAPTER XVII.

A Psychological View of the State of Guiteau.

Charles Guiteau, who assassinated President Garfield on the 2nd of July, 1881, presents the following family history for our examination and reflection : An aunt of the prisoner died while laboring under senile dementia ; her daughter — his cousin — became a victim to religious depression in her sixteenth year; her sister — another cousin — was deformed ; another aunt died insane ; the son of this aunt died in acute mania; an uncle of Guiteau's was weak-minded and destitute of self-control, and the son of this uncle died insane. Guiteau's mother suffered from ill health before his birth ; he had a deformed sister, and his father was mentally unsound through his whole life, as he entertained untenable views respecting religion, and during the last six weeks of his life he was thoroughly an insane man, declaring that his fatal illness was unnecessary as a passage to eternity, as he had already realized that state. How, with such a family history of insanity and predisposition, could any man escape mental unsoundness and morbid exaltation ? Was it possible for his original mental constitution to be normal? His idea in conceiving himself the instrument of Divine vengeance or inspired, that he was the instrument or agent of Deity, is very similar in its nature to his father's insane delusion that he had attained immortality by his present union with the Saviour. His family, as far back as 1876, had entertained strong suspicions as to Guiteau's mental responsibility. He had lectured absurdly on the Second Advent, left another assembly,

23

in a very irrational way, in the midst of an address, and, armed subsequently with an axe, he threatened the life of his sister. Dr. Rice, in 1876, when called by the family to examine Guiteau, pronounced him *insane*, basing his opinion on the following grounds, viz.: hereditary influence and exaltation of his whole emotional nature, and says: "This exaltation was attended with explosions of emotional feeling which appeared to arise *from centric causes, not from eccentric causes ;*" or, in other words, it was an emotional storm with no adequate external cause. Just such scenes are of daily occurrence in an insane asylum. Dr. Rice also detected incoherence of thought and an excessive egotism, and also an intense pseudo-religious feeling. Dr. Rice said that Guiteau was always talking about Christ and Christianity and religion, without having become impressed, in Dr. Rice's opinion, with any of the moral principles of Christianity. He also discovered weakening of the judgment and some impairment of the intellectual faculties. He did not discover very much disturbance of the intellectual or of the perceptional powers, and was unable to discover either illusion, hallucination or delusion. He found a strong hereditary predisposition and a congenital moral defect, a true moral imbecility, such as exists in two cases now within the writer's personal knowledge. Puberty added to the emotional exaltation, as it naturally would, and Dr. Rice's very wise opinion was, that Guiteau, in 1876, *was dangerous and perhaps incurable*, and advised his seclusion in an asylum. He was right again, *for these are not curable cases*, in my opinion, and the prognois is bad in every one of them, and they all belong in asylums for the chronic insane, where they should be kept for life.

Guiteau, hearing of the determination to put him in an asylum, suddenly disappeared, and unfortunately this consummation was not reached. If he had been then incarcerated, I think any intelligent superintendent of an insane asylum in which he was a patient would have detected, without difficulty, sufficient disorder of the cerebro-psychic functions to constitute unequivocal insanity.

He had shown peculiarities and wandering in public speaking; had made grotesque mistakes in gardening; had been requested by a Miss Lockwood to leave her house, as he had behaved in such an eccentric manner; had given an insane lecture relative to Boston and two-thirds of mankind going to perdition; had called upon Senator Logan in summer dress, with low shoes, when snow was on the ground; had been in the habit, according to Mr. Hubbard, of gesticulating wildly, vehemently and incoherently, or sitting moodily in a corner; did not speak, according to Mrs. Scoville, until he was six years old, and neglected his studies to investigate the mysteries of the Shakers and subsequently became so violent in temper and conduct that he had to leave her house; that he was, according to the Hon. Emery Storrs, in a state of exaltation, and that he had an ill-balanced mind, ill-balanced judgment, and an utter want of average good common sense; that he had a conversation with Geo. T. Burrows, in which he talked incessantly of the second coming of Christ in a very excited manner, and that from his whole deportment Mr. Burrows thought him a fool or crazy; that in the Oneida community, according to the manager Joscelyn, he was the most egotistical of men, declared himself the leader and was a decided fanatic; that he made love to Mrs. Parker's daughter, a girl of 13 years; that he announced that he was to marry a lady worth a million, and that together they could ably represent the American people in Vienna; that he asserted that he was predestined to be the agent of the Diety in the homicide of the President, and that he had no volition in the matter; that his book called "Truth" was a second gospel dictated by inspiration, and that his convictions and impulses proceeded from God; that he said that he felt no compunction or remorse since the removal of the President, and was satisfied that he had conferred an incalculable benefit to his countrymen; that he was cool and com-

posed when the President fell bleeding, and slept soundly
on the night of the homicide, a fact in itself strongly
indicative of insanity ; that by the expectation of supernatural aid he not only meditated but undertook absurd
and impracticable schemes and speculations which were
alike disproportioned to his abilities and resources ; that
he showed presumptous interference with and unfounded
pretensions concerning the political party he connected
himself with, and that finally he showed a decidedly
insane behavior in court when on trial for his life, in that
awful crisis showing that he had a mind so incoherent as
not to be able to understand the seriousness of the crime
he had committed, but in his insane egoism rather
delighting in being the most prominent and notorious
person for the time being in the whole country. What
else but insanity could be expected from a child born of
a diseased mother and an insane father, and steeped
during childhood with fanatical religious views? Upon
the trial, Guiteau's counsel submitted the following
hypothetical question to all the experts examined :
Assumed that there is in the blood of Guiteau a strong
hereditary taint ; that at the age of 35 years his mind
was so much deranged that, according to Dr. Rice, he
was a fit subject to be sent to an asylum ; that frequently after that date, during the succeeding five years,
he manifested such decided symptoms of insanity, without simulation, that many different persons conversing
with him and observing his conduct, believed him to be
insane ; and further, that in or about the month of June,
1881, at or about the expiration of said term of five years,
he became demented by the idea that he was inspired of
God to remove by death the President of the United
States ; that he acted on what he believed to be such
inspiration, and on what he believed in accordance with
the Divine will in the preparation for and in the accomplishment of such a purpose ; that he committed the act
of shooting the President under what he believed to be
a Divine command, which he was not at liberty to

disobey, and which belief made out a conviction which controlled his conscience, and overpowered his will as to that act, so that he could not resist the mental pressure upon him ; and lastly, immediately after the shooting, he appeared calm and as if relieved by the performance of a great duty, and that there was no other adequate motive for the act than the conviction that he was executing the Divine will for the good of his country.

Dr. Nichols, the superintendent of the Bloomingdale Asylum for the Insane, New York, guided by this hypothetical statement and from his personal examination of the prisoner, testified that he considered him insane.

Dr. Norton Folsom, of Boston, the professor of psychology in Harvard University, accepting the hypothetical question to be correct and trustworthy, pronounced the prisoner to be unquestionably insane.

Dr. Godding, of the Government Hospital for the Insane at Washington, taking for granted that the propositions upon which the opinion of preceding medical witnesses were based, were at once accurate and scientific, but without entertaining any convictions as to the truth or falsity, was of opinion that the person described therein was unquestionably insane.

Dr. McBride, of the Milwaukee asylum, concurred with the view of Dr. Godding.

Dr. Walter Channing, of Brookline, Mass., who served for eight years in the asylum for insane criminals at Auburn, N. Y., said, relying upon the truth of the facts set forth in the hypothetical question, he would declare Guiteau insane.

Dr. Fisher, for years connected with the lunatic asylum at South Boston, on the hypothetical question, although unwilling to rely on such restricted evidence, would regard Guiteau as of unsound mind.

Dr. Spitzka, of New York, entertained the belief that Guiteau was insane at the time of the committing of the crime, and had always been of unsound mind. Thought Guiteau a moral imbecile.

Dr. Noble, the jail physician, whose opinions had great weight with the jury, regarded Guiteau as sane, bright, and intelligent.

Dr. A. McLane Hamilton thought Guiteau sane, but eccentric.

Dr. Worcester believed Guiteau to be sane when the deed was committed.

Dr. Talcott, superintendent of the State Homœpathic Hospital for the Insane at Middletown, from his examination of Guiteau in jail, from his manner and a review of the events of his life, would infer that he was sane.

Dr. Stevens, of the Retreat at Hartford, had examined Guiteau four times in prison. He adopted the hypothetical propositions of the prosecution and regarded the prisoner as sane, but acknowledged that the tendency to insanity was more than ordinarily strong in his progenitors. He observed certain indications of abnormality in the state of the pupils, the tongue, the head, and slightly in his articulation, but thought that otherwise Guiteau was healthy.

Dr. Jamin Strong, superintendent of the asylum at Cleveland, Ohio, thought that if the hypothetical propositions presented by the defense were trustworthy, Guiteau was of unsound mind.

Dr. Shew, superintendent Middletown (Conn.) Hospital for Insane, from his examination of Guiteau, regarded him as sane, but confessed that egotism was a marked symptom in the criminal insane. Thought that in court Guiteau did not feign but acted under the influence of his natural disposition and manner.

Dr. Evarts thought Guiteau sane.

Dr. Macdonald, superintendent of the New York City Asylum for the Insane, did not recognize moral insanity as a disease but as another name for wickedness, and believes Guiteau had feigned in court what he (Guiteau) believed to be insanity. Believed the prisoner to be sane.

Dr. Barksdale, of Virginia, thought Guiteau acted insanity in court, and was a sane man.

Dr. Collender, of the Tennesee asylum, believed Guiteau to be of sane mind, on the ground that his replies were satisfactory and intelligent.

Dr. Walter Kempster thought Guiteau of sound mind, but admitted that lunatics could often curb and control their unhealthy tendencies and designs till at length the morbid influence controlled them.

Dr. Gray, superintendent of the New York State Lunatic Asylum at Utica, testified that he did not believe in moral insanity, and regarded Guiteau as of sound mind. He said that such self-control, self-direction and self-guidance as he displayed was antagonistic to anything that he had ever seen in his personal experience with the insane.

Guiteau's every mental process which resulted in the establishment of a motive towards the act of homicide, *was a diseased mental process.*

Suppose Guiteau's brain or his mental processes had originated the idea of a motive to suicide instead of homicide. In the face of his family history and his own history, as I have given it, what would have been the verdict of the eminent gentlemen who pronounced Guiteau a sane man? I think the unanimous verdict would have been — mental unsoundness, inciting to suicide.

Again, it has been proved that "with a regularity which possesses all the significance of law, the maximum of deaths thus caused (by suicide) is coincident with the months of ending spring and commencing summer ; that is, at these periods of the year, when change from one season to another takes place, the numbers show appreciable increase compared with those of fixed seasons." Guiteau, again, had damaged a naturally weak mind by excessive and laborious thinking out of questions raised in connection with religion, and it would have been a very natural outcome of all this, if, instead of exaltation, he had displayed deep depression and had committed suicide, as a great many persons do by reason of the exaggerated inward struggles of the conscience in weak minds susceptible to morbid impression from their weakness.

Respecting the question of moral insanity, which need not have been introduced into the trial at all, and the existence of which Dr. Gray and Dr. Macdonald denied entirely, we would simply say, that we think these eminent men "suffered their judgments to become biased by the idea that the faculties of the mind cannot act separately and that to derange one must necessarily and appreciably disorder others." Because the emotions are affected by disease, it does not necessarily follow at all that the intellect is, and clinically, any man who has seen much insanity and closely studied it, will have wit- nessed many cases in which the insanity has not resulted from the perversion of reason by disease. Dr. Gray, although he stated on the trial that kleptomania was thieving, dipsomania was drunkenness and pyromania a crime, incendiarism ; would not wish, I think, to be understood as denying the fact that kleptomania,* pyro- mania, dipsomania, and homicidal and suicidal impulses, may depend on morbid states of the reproductive system, upon uterine disorder in women, and that in these cases we find a form of mental derangement in which practical psychiatry demonstrates that it is the feelings and emotions rather than the reasoning processes which are disordered. Rush, Pinel, Pritchard, Maudsley, Bucknill and Tuke, and the late Dr. Ray have, in their writings and teach- ings, all inculcated the doctrine that mental disease may attack and derange the affections, the emotions and the will while the intellect or reason remains practically

---

*See case of Madame de Kouvitchinski, of an eminent and wealthy Russian family, arraigned for theft before a police court in Paris, in 1877, where the eminent Russian specialist in insanity, Dr Frabrinus, declared her unequivocally insane. "London Journal of Psychological Medicine," for October, '81, p. 256.

intact and not appreciably affected. Practical psychiatry also demonstrates that not unfrequently the intellect, in many of these cases, becomes finally involved, and this, I think, would be the case with Guiteau very possibly, should he live long enough. Theoretical views and metaphysical conceptions of mind will never harmonize with the facts which psychiatry demonstrates, and we have to deal with and acknowledge *facts*, as psychologists, and should do so fairly and impartially.

In the Guiteau case, on the one side, the evidence was supposed to show a chronic condition of insanity before the crime; and, on the other side, to show an exceptionally quick intelligence and decided powers of discrimination. I think the evidence on both sides to have been correct, and also think that very clearly Guiteau has for years been the subject of emotional insanity, taking an exalted form and being characterized by *exaltation* regarding religion, pride, vanity and ambition. The intellect has been intact, while the feelings and the moral sentiments have been affected by disease: the emotional insanity finally taking on a destructive character, as it not unfrequently does when the propensities, instincts, or desires are involved, homicidal or suicidal mania being the form of insanity, as the case may be. His case comes under the head of the psychical degenerative states affecting the brain injured by hereditary vices of conformation or nutrition. To convey a correct idea of what I believe in Guiteau's morbid mental state to have been, is, I fear, somewhat difficult. Unless at the last Guiteau thought himself the agent of God to accomplish the "removal" of the president, I believe him to be a man who, while he has had no decided delusion, has been influenced by the most exalted notions respecting himself: his every gesture, and expression, and conversation displaying his diseased self-love. He has exhibited the *monomanie raniteuse*, associated, however, with great intelligence, and an extent of knowledge of which many sane men in

the court room at his trial might be justly proud.
In the court room he exhibited the excitement, susceptibility and fury of a monomaniac, and it is
somewhat surprising that this fact seemed not to be
recognized by the many able alienists present. It is a
classical fact that monomoniacs have a general sense of
well-being, and seize on the cheerful side of everything;
that they are satisfied with themselves, and are content
with others; that they are controlled by vanity, and self-
love, and delight in their own vain-glorious convictions;
that they are susceptible and irritable; that their determinations are violent; that they are inexhaustible in
their loquacity; that they dislike opposition and restraint and easily become angry and very furious. Is
not this a picture of Guiteau on trial for his life? He
has exhibited a religious exaltation amounting to Theomania. This is a much rarer mental state than religious
melancholia or depression, but it is a state distinctly
recognized by alienists, and Guiteau has clearly exhibited
it, and it would have been at all improbable for Guiteau,
if he had lived long enough, to have manifested the opposite extreme of religious melancholy or depression. In the
development of insanity generally, melancholy precedes
mania, but these cases, of which Guiteau is an example,
are exceptions to the rule. Guiteau was never, I think,
very different from what he was on trial. He never
evinced any consciousness of ever having done anything
wrong, and was so completely destitute of shame or remorse, and proved himself so utterly incorrigible throughout life, that I can only satisfy my doubts by pronouncing him insane. He had been a scourge to his family
from childhood, and on him little moral influence could be
exerted. He appeared to be quite destitute of the moral
feelings and without human kindness. I beleive that
there is a certain class of the insane in whom we find the
union of intellectual ability with congenitally feeble
moral powers, a moral insensibility and necessarily a
proportionate irresponsibility. I do not understand how

any alienist of experience can have failed to recognize such cases, if he be a man of any discernment. I know of a case to-day, in which the intellectual faculties are not only equal to, but far superior to the average, and where there is, I believe, a perfect moral insensibility. I believe the disease in Guiteau's case to be a congenital deficiency, the result probably of perverted nutrition of the textures of his brain, occurring, perhaps, even during fœtal life. I believe that his brain had undergone pathological changes, which induced defective moral power, and think he would have been regarded as insane rather than criminal, if his acts had not made him decidedly amenable to the laws of his country. In the former case he would have been a life-long inmate of an asylum, while as it was he was consigned to the gallows. While Guiteau was quick, had an excellent memory, and could acquire knowledge easily, and although in the abstract he knew the difference between right and wrong, as do most of the insane, yet he had appeared utterly incapable of following the former like other men. The form of insanity is emotional insanity proper, with partial exaltation or exalted emotional condition. It is *reasoning mania*, as the intellect is intact. He had not been deprived of the use of his reason, but his affections and disposition were perverted, and the case was one of general moral obliquity dependent upon cerebro-mental disease. In the last act of Guiteau's life drama, the *ego* may have been overborne by an impulse not sudden, but which the will had no longer power to restrain, as the result of maniacal excitement. Had Guiteau lived long enough I should have thought the prognosis of his case very gloomy, and should have expected to see him end in mania or dementia. Finally, I consider that had the element of melancholy or depression predominated in the mental state of Guiteau, rather than that of exaltation, as it might readily have done, the nation would then have been spared the profound grief with which it has been stricken, and the last act in the drama of Guiteau's life would have been a suicide instead of a homicide.

SCOVILLE AND HIS WIFE.

In the Scoville insanity case the court room was crowded. Mr. Scoville testified to traces of insanity in the Guiteau family from the grandfather down. Mrs. Scoville, he said, first showed signs of insanity when their son became sick. She fell in love with the physician attending him and confessed it. Subsequently she denied she had made the confession. She soon became violent, nervous, and irritating, and attempted to leave the house. She confided all her griefs to the servants, lost affection for her family, except for himself, and within the past six months had deserted him, too. He related in detail his recent well-known troubles with her. She formed a friendship with George W. Earlie, an alleged newspaper man, and confessed to witness that she loved Earlie. They wrote letters addressed to Mrs. Garfield and others and sold them to the newspapers. Mrs. Scoville was very intimate with George. She was very sly now, although formerly frank. She was lately changeable and fickle, and often provoked him.

At the afternoon session Mr. Scoville continued his testimony. He said he believed Mrs. Scoville to be insane. She was sometimes irritable and nervous. She had said if Guiteau was taken away her coffin should be got ready, and that if she went her daughter Bertha should not remain. She said she was afraid to live in the same house with witness, as if he should die suddenly people would believe she had poisoned him. She showed loss of feeling and indifference to sentiments which affect ordinary people. The execution of Guiteau had little effect on her.

In the cross-examination Scoville asserted that he had always been a kind and devoted husband, and that all his efforts were directed to restrain his wife from going about and into gentlemen's offices, and making acquaintances in irregular ways. He had never accused her of improper intimacy with men. She had an idea that she was growing younger, and was in the habit of undressing and sleeping in her room with the door open, an impropriety concerning which he had remonstrated with her.

The next witness was Mrs. Fannie Scoville Harper, Mrs. Scoville's daughter. Her testimony was as to the queer actions and epileptic fits of her mother since witness was ten years old. The evidence, in the main, corroborated that of her father.

THE SCOVILLE FAMILY TROUBLES.

In the Scoville insanity case, Fanny Scoville Harper, the daughter, testified that she had intercepted a note from George W. Earlie to Mrs. Scoville, the contents of which were very improper. She had seen her mother kiss three men whom she had no right to kiss, and had heard of many others whom her mother had kissed, but she declined to give their names. Dr. A. McFarlane, of Jacksonville, Ill., who was subpœnaed, but did not testify, in the Guiteau murder trial, and who has been for twenty-six years in hospitals for the insane in New Hampshire and Illinois, said, eighteen years ago Luther W. Guiteau brought to him an insane sister for treatment. From Guiteau's talk, which was very peculiar, he set him down as a crazy man. The impression was very strong on the witness, and he readily saw how Luther's son could get into a state of mind necessary to kill President Garfield. He had met Mrs. Scoville at Washington last winter. She had a fierce controversy with John W. Guiteau about the witness' testimony. John declared that nobody should testify that his father was insane. Mrs. Scoville said witness should testify, and declared that all the family were predisposed to insanity. He believed that Mrs. Scoville's fainting and epilepsy were sure to produce insanity. Her being a chaste and loving wife until recently and her sudden change was an evidence to him of insanity. He held, in short, that the whole family, including John W., were of unsound mind and insane.

Another very interesting case, although not quite analogous to that of Guiteau, is that of Roderick Maclean, who was tried at Reading, England, April 19, 1882, the indictment charging him with traitorously and maliciously compassing the death of her Majesty the Queen of England, and with having, on the 2nd of March, discharged a pistol, loaded with powder and bullet, at her majesty, in the parish of Windsor. The prisoner said he committed the crime under the influence of a condition of mind brought upon him by the Almighty. Dr. Mandsley, in 1854, after an examination, expressed the opinion that Maclean was not of perfectly sound mind, and recommended that he be secluded. During the year 1880 he wrote insane letters to his sister, showing clearly that he had inclinations towards homicidal mania. He had a high idea of his histrionic abilities and thought himself a great actor. He differed from Guiteau in being more markedly insane, with actual delusions, and in being less of a typical case of reasoning mania where the intellect is merely intact. Dr. Alfred Goodrich examined him in 1873-4, and advised his careful supervision to prevent his doing any injury to himself or others. Dr. Hitchins, of Westminister, on the 3rd of June, 1880, also came to the same conclusion, and Maclean was then confined in the Somerset Lunatic Asylum and suffered from homicidal mania. He was discharged *recovered* (?) on February 21. He had a delusion while in the asylum that he must kill someone. Dr. Edgar Shephard, formerly superintendent of Colney Hatch Asylum for twenty years, testified that he regarded the prisoner on the 24th of March and on the 10th of April, 1882, as of unsound mind unquestionably. Dr. Shephard said : " I should say that the prisoner has very marked congenital defects, which handicap him very heavily ; he has a very narrow head, with the high arched skull so commonly associated with idiocy and insanity ; he is not a man who could reach a fair standard

of moral or physical health ; he has a nervous hesitancy
of speech amounting to a stutter—imperfect vocal articulation I should call it ; he has a scar on the right side
of the head, about two inches long, the result of an accident about thirteen years ago, as I understand it; this
scar is very tender on pressure ; he complained to me of
a shooting pain through the forehead ; I found that he
had delusions of an unmistakable character; he said
persons in *blue* were against him and always had been ;
that he had a mysterious connection with the number
four and "the blue," and this combination of figures
was always disasterous to him.   He told me a few weeks
ago he went to Somerset House in the Strand, to ascertain whether he was registered or baptized.   Finding he
was neither the one nor the other, he thought himself
more injured than ever, and he determined to bring his
case under observation by taking the step he had done.
I pointed out to him the inadequacy of his grievance to
the measures adopted for redress, but he did not seem to
see it at all.   He had a perfect right to do what he had
done, because it had been revealed to him in early life
that he had a great secret power over all mankind ; he
also said that he was related to the royal family as
much as Geo. IV, and the crowd would have torn him to
pieces the other day had it not been for Jesus."   Dr.
Shepherd regarded him as an imbecile, and thought he
had always been so, with a delusional mania, and that
the real question of right and wrong did not present
itself to such men.   After much more similar testimony,
Dr. Wm. Orange also testifying, the Lord Chief Justice
having reviewed the evidence adduced, thus summed up
the case to the jury :   " It is for you to say whether you
consider the prisoner to be not guilty on the ground of insanity ; but if you find him not guilty, you should be careful to add the words ' upon the grounds of insanity,' and
for this reason, that if you find him not guilty without such
a qualification, he would be entitled to leave the box, and
would perhaps repeat his crime ; but if you add that, then

by the statute he passes into the control of the government, and will be locked up, as it is called, during her majesty's pleasure, or until such time as those who should advise the crown are satisfied that he can be safely let free. It is a merciful verdict; it saves the man's life, as it ought to be saved, if he is not a moral agent, while on the other hand, it protects society against a repetition of these outrages, because he will be placed under the control of the government and will be kept in custody as long as it will be right for him to be kept. I do not desire to say one word about the nature of the act if the man had been responsible for it, because it seems to me that burden of proof has, at all events, been very largely met by the prisoner; it is, of course, for you to say whether that burden has not been fully satisfied by him; but if you are convinced that he did the act when he was responsible, no words could be too strong, no punishment too heavy for him. If he was not responsible for it, although the life which he put in danger was a life inestimably precious, he ought to be protected as much as if he had only committed the most trivial offence against the meanest subject of the realm. Gentlemen, you must now consider your verdict and say whether you find him guilty or not guilty on the ground of insanity."

The jury retired at twenty-three minutes past five o'clock, and after an absence of five minutes returned into court. Their verdict was not guilty on the ground of insanity. The usual order was then made that the prisoner be detained in strict custody during her majesty's pleasure, and the prisoner, who manifested no emotion, was removed from the dock.

At the autopsy held upon the body of Charles J. Guiteau, three quarters of an hour after death, under the direction of Dr. Lamb, U. S. A., by Wm. J. Morton, M. D., and C. L. Dana, M. D., the following appearances were observed. This I extract from the "Journal of Mental and Nervous Diseases" of July, '82, the able and accomplished editor of which has laid the profession

under obligations for his indefatigable labors in behalf
of science in this case. "The most noticeable asymmetry
was a slight flattening of the upper and anterior part
of the right parietal bone. The flattening ended sharply
at the coronal suture. It included a space about half the
size of the palm of the hand, etc." *Brain membranes.*—
The dura mater was quite strongly adherent in places to
the inner surface of the skull. Near the trunks of the
middle meningeal arteries upon each side, the membrane
was thickened and strongly adherent to the bone, though
it could be stripped clean. It was also adherent near the
longitudinal sinus in front. There was at these points,
probably, a slight chronic pachymeningitis externa.
There was no exudation upon the inner surface of the
dura, anywhere. The cerebral sinus contained but little
blood. There was rather more than the average amount
of pacchyonian granulations distributed along the middle part of the upper surface. *Arachnoid.*—There were
very well marked milky opacities of the arachnoid,
extending over the upper convex surface. These opacities were over the fissures only ; in some parts they had
a somewhat yellowish look. The sub-arachnoid space
contained no abnormal appearance. The *Pia Mater*
presented no abnormal appearance ; it came off easily from
the brain. The *blood-vessels* of the membrane were not
full, and the general appearance of the brain was anæ-
mic." The brain weighed 49 1-2 ounes ; its consistence
was normal ; there was no apparent asymmetry of the
two hemispheres. "The cerebellum was well covered ;
the occipital lobes were not noticeably blunt or sharp ;
the frontal lobes were peculiarly shaped. Looking
at them from in front and above, they presented two pro-
truding points from which the surface sloped away in a
concave curve. This pointed apex of the lobe, with the
concavity of the orbital and the beginning of the frontal
surface, was carefully noted by all of us at the first exposure and removal of the brain." From the examination
of the lobes and convolution it was determined that, "on
24

the whole it would appear (1) that the brain was marked
by an unusual number of cross and secondary fissures,
especially in the frontal lobes ; (2) that it was not of the
confluent fissure type ; (3) that the convolutions on the
two hemispheres were quite asymmetrical." Respecting
the interior of the brain, "the white substance was
somewhat whiter than usual and of normal consistency.
The grey cortex was measured and seemed to be somewhat thinner than usual, and of normal consistency.
Eight or more measurements gave a thickness varying
between 1-8, 1-9, 1-12 and 1-16 of an inch. The
ventricles were dry, the ependyma normal, and the
choroid plexus showed nothing noticeable. No spots
of hemorrhage or softening were found, and no tumors
were present." The microscopic examination of
Guiteau's brain, made by Drs. J. W. S. Arnold, E. O.
Shakespeare and J. C. McConnell, exhibited changes in
its constitution, denoting initial *dementia paralytica* (progressive general paralysis) of the insane. In sections of
the corpus striatum were found decidedly abnormal blood
vessels, "particularly capillaries and venules." "Their
perivascular lymph spaces were often more or less completely filled with masses of yellowish brown pigment
granules which appeared to be the degenerated remains
of old blood extravasations." In areas very numerous,
but mainly limited to the gray or ganglionic substance,
the capillary blood vessels presented their walls in a state
of granular degeneration. Sometimes these granules
were limited within the endothelial cells constituting the
wall of the capillaries, but often they were found for a considerable distance completely encircling the vessel. In
the gray or ganglionic matter of these sections were quite
numerous areas in which alteration of the neuroglia and
of the ganglionic nerve corpuscles were very plainly visible. In them the pericellular lymph spaces were much
crowded with lymphoid elements." (This is a condition
which the writer of this volume has for years insisted

upon as a frequent cause of the nutritive defect which results in chronic insanity, these lymphoid elements tending to undergo a fibroid metamorphosis.) " In some areas the whole space seemed to be crowded by collections of such cells, no trace of the neuroglia cell or nerve corpuscle remaining. Most frequently, however, neither the encompassed nerve corpuscle nor the neuroglia cell was destroyed." (My remarks with regard to these aggregations of white cells, the reader will find at pages 141-142 of my "Manual of Psychological Medicine).''

Respecting the cerebral cortex "in the second, fourth and fifth layers the blood vessels presented, in a marked degree, degeneration similar to those remarked in the corpus striatum. In the second, fourth and fifth layer the pericellular spaces, both of the neuroglia cells and of the ganglionic corpuscles, were more or less filled with lymphoid cells. In these layers, some ganglion nerve cells were also quite freely pigmented, etc., etc.

"In conclusion, your committee have no hesitation whatever in affirming the existence of unquestionable evidence of decided chronic disease of the minute blood vessels in numerous minute diffused areas, accompanied by alteration of the cellular elements, in the specimens of brain submitted for the examination. While the lesions found were most markedly in the corpus striatum and in the frontal region of the cerebral cortex, yet they very diffusely pervaded all portions of the brain which the sections represented. etc., etc. They have not been called upon to pass upon the bearing the lesion found might. have upon the state of the subject's mind, and therefore do not offer an opinion."

The deductions I draw from the autopsy and microscopical examination is that confirmatory of my opinion, written for the " London Journal of Psychological Medicine and Mental Pathology,'' and expressed publicly in the New York Medico-Legal Society, before Guiteau's death, viz., that he was an insane man, who would have

died demented, in all probability, had he lived long enough
for the natural history of his disease to have gone through
its regular evolutionary periods.   Both the clinical history and the pathological and morbid histological findings
of this case bear me out in this assertion.

CHAPTER XVIII.

PERSONAL IDENTITY IN MURDER CASES.

DR. A. WALTER SUITER has kindly given me some points of medico-legal interest in the scientific investigation of the case of The People v. Roxalana Druse, together with a statement of specimens representing the *corpus delicti*.

I invite your attention to some points of medico-legal interest in the history and prosecution of a case, which in some respects is one of the most interesting, unique and peculiar that can be cited from the annals of criminal jurisprudence; and which, because of certain legal questions which have been raised in connection therewith, has become somewhat notorious throughout the State. So far as some features of the scientific study of it are concerned, it is quite unparalleled in the history of crime.

I refer to the case of The People v. Roxalana Druse, who was indicted for the murder of her husband, William Druse, at the April term of the Court of Oyer and Terminer, held in the county of Herkimer, in the year 1885, and tried at an adjourned term of that court, in September, 1885 : was convicted of murder in the first degree, and was executed February 28, 1887.

It will be necessary, in order to make intelligible what remarks I desire to make, to ask your indulgence while I recount briefly the details of the early history of the crime. The parties resided upon a farm in the town of Warren, Herkimer county, the family consisting of William Druse, his wife, a daughter aged eighteen, a son aged ten, and a nephew aged fourteen, who temporarily resided with them.

Because of certain suspicious acts and circumstances, the near neighbors were aroused to the belief that something wrong had taken place upon the premises. The sudden disappearance and continued absence of the old gentleman, who was the head of the family, coupled with a knowledge of domestic infelicity, served to add confirmation to their suspicions, and led them to institute something of an inquiry. By some shrewd amateur detective work on the part of these neighbors, a partial confession was obtained from the lad Gates, which eventually brought the matter to the attention of the authorities, and made it almost beyond question that a murder had been committed in a most inhuman and brutal manner, and that an attempt had been made to dispose of the body by a novel and ingenious method — by such means, as subsequent revelations proved, as the courts are seldom called upon to consider.

A coroner's jury was organized on January 19th, 1885, at which time Dr. Suiter was requested by the district attorney of the county to visit the scene of the tragedy, and directed to search for any evidence of a scientific character which might be of assistance in the investigation.

From the evidence taken before the coroner it appeared that the manner in which the deed was committed was substantially as follows:

After an altercation between the husband and wife at the breakfast table on the morning of the tragedy, which was the 18th of December, 1884, about one month previous to its discovery, the wife called the two boys and directed them to go from the house on some pretext, and proceeded, with the assistance of her daughter, to the commission of the crime. Approaching her father, who was still sitting at the breakfast table, the daughter threw a rope about his neck, and thus holding him fast, her mother fired a shot from a revolver in the direction of the man's head. It is said that the ball entered the back of his neck. At this time, probably desirous of implicating the only persons who might cause the crime to become known, the mother opened the door and called the boys. When they had entered the house, she closed and locked the door behind them. Handing the revolver to the older boy, she directed him to fire it into the man's body, threatening at the same time to shoot him if he did not comply. The terrified boy did as he was told, and the man fell backwards from the chair to the floor, bleeding profusely from his wounds. Observing that his wounds were apparently not of a fatal character, the woman sent one of the boys for an axe, with which she deliberately chopped his head from his body, carried it into an adjoining room, and placed it in the stove, after which they placed the headless body upon a straw-tick, and carried it also to the adjoining room. The body was then cut into convenient sized pieces, which were successively deposited in the stove, during which time a very hot fire was kept up by means of dry pine shingles which had been procured for the purpose.

When the work of incineration was thought to be complete, the ashes were removed from the bottom of the stove, placed in a bag, and carried to a swamp about one half mile distant from the house, and deposited under a clump of bushes; from which place they were subsequently recovered, and put in Dr. Suiter's possession for examination — a chaotic mass of finely pulverized wood and bone ash, bits of charcoal, and small fragments of bone of irregular form, which could not be recognized as parts of a human skeleton, without careful and diligent study.

I shall now endeavor to interest you by making reference, as briefly as possible, to those points which I deem important and worthy of notice in the case, and will exhibit to you all those parts of the body of the deceased which it was possible to recognize and identify.

It is obvious that the principal question in dispute in such a case, would naturally be with reference to the establishment of the fact, in the first place, that a murder had taken place ; as proof of death, which generally comprehends the discovery and identification of a body in any case of alleged murder, has always been regarded as essential to constitute the substance of the charge, or *corpus delicti*, in the prosecution.

Lord Hale, who is quoted by nearly all legal and medico-legal works on this point, says : "I would never "convict any person of murder or manslaughter unless "the fact were proved to be done, or at least the body "found dead." "*De corpore interfecti neccesse est, ut constet*" is an ancient legal maxim, and many cases might be cited to illustrate the grevious errors which may be made by the non-observance of this almost universal rule.   Conviction upon the testimony of eye-witnesses, with strong and indisputable corroboration, are the only exceptions, and the exceptions are extremely rare.

This question was made the basis of the defense in this case, the theory being that the prosecution would fail to show that William Druse had been killed, and that the small fragments of bone could not be identified as parts of a human skeleton, and that they could not be distinguished from fragments of the bones of the lower animals.

Very few cases of a character similar to the one under consideration as to the identification of the body are reported in the various works on medical jurisprudence, where such a small part of the body was recovered, and especially where an attempt had been made to dispose of it by fire.   I venture to state that in this respect this case must stand almost, if not quite, without parallel in the history of crime.   One which bears the nearest relation to it was the celebrated Webster-Parkman case, which doubtless many will readily call to memory. That occurred in the year 1849, and was then con-

sidered of so much importance in the particular respect
to which I refer, that volumes of literary matter were
published at the time and are preserved for reference.
Probably no case was ever more extensively quoted by
medico-legal writers in all parts of the world, as to the
question of *corpus delicti.*

An attempt was made by Dr. Webster, a professor of
chemistry in the Massachusetts Medical College, to dispose of the remains of his victim in a coal furnace.
Identification was only established by the recovery, from
the slag of the furnace, of a portion of a plate of false
teeth, which was subsequently recognized by the dentist
who made it. Conviction and execution of the prisoner
followed. And yet, in that instance, instead of a few
small fragments of calcined bones found in connection
with a quantity of ashes, charcoal, etc., a considerable
portion of the body was recovered. A number of eminent medical gentlemen were engaged in the investigation
of the scientific evidence, and succeeded by diligent
search of the ashes from the furnace, in positively iden-
tifying bones and fragments representing nearly every
region of the body.

Appleton Morgan, Esq., in an article on "Personal
Identity," published in 1877, makes the following observations regarding the Webster case and the question of
the establishment of the *corpus delicti :* "I cannot refrain
from remarking that it might not be well to place too
much reliance on the Webster case to-day. Had Dr.
Webster been tried in 1876 instead of 1850, upon the
same evidence, it is very probable that he would have
been acquitted.

"Courts hold very strictly that the *corpus delicti* must
be shown beyond any reasonable doubt. Even in the
Webster case, which, with the single exception of the
Gardelle case, must so far stand unparalleled as the only
case where fragments of a human body, unrecognizable
by ordinary investigators, were actually identified after
passing through an intense chemical heat intended to

utterly destroy them, all recognition would probably
have failed, had it not been for certain artificial matter,
viz., the gold introduced into the murdered man's teeth
by a dental surgeon."

I make this quotation for the double purpose of illustrating the necessity of absolute proof of the death of an
individual, as essential to conviction for the crime of
murder, and to call attention to the fact that, in the case
under consideration, the prisoner, Mrs. Druse, has been
convicted, and will without doubt be executed, upon very
much slighter but just as certain evidence of the discovery
of a body, as you will presently see, than in any other
case so far recorded.

The examination of the materials presented was conducted as follows, and with the following results :

It was thought necessary, in order to facilitate the
examination of the fragments of bone, to pass the materials through sieves of different sized meshes. This was
accordingly done.

All the large pieces of bone were then selected out from
the coarsest of the siftings, by means of a fine forceps,
and were carefully examined, one by one, and prominent
points upon them were compared with prominent points,
processes, foramina, etc., which they seemed to resemble
upon the human skeleton, and also those of all the lower
animals which would be likely to be found about a farm
house in the vicinity, with the result of fixing the identity
of a small number of fragments, which without doubt
could be positively declared to be human. These,
together with the coarser unidentified fragments, were
then carefully weighed, and the weight compared with
that of a medium sized human skeleton.

By a quantitative chemical analysis of the finely sifted
ashes, the bone-ash was separated from the wood-ash,
and the weight of the bone-ash, as computed, added to
the weight of the fragments, by which it was found that
a little more than two-thirds of the entire skeleton had
been recovered, but proving beyond question that a con-

siderable portion of the body had been disposed of by other means or at some other time. This observation was verified.

After recovering all the evidence from the ashes which could possibly be identified, and positively sworn to, as parts of a human skeleton, the bits of bone were classified, catalogued, and fixed upon plates of glass, in order that they might be viewed both anteriorly and posteriorly, and thus they were presented to the court and jury.

Another very important question which was raised by counsel for the defense, interesting because of the paucity of recorded facts relating to the subject to which it refers, was in regard to the alleged cremation of the body.

It would be difficult to find among professionals or laymen, any persons who, without special and experimental observation, could give anything like an accurate idea as to the length of time it would take to completely consume by fire, beyond recognition, a human body in a common wood stove; or as to the amount of fuel which would be required to perform such an operation.

In all the noted text books on medical jurisprudence to which reference can be made, but a very few lines are, as a general thing, devoted to the treatment of the subject of cremation of bodies as a means of concealing crime. A few cases are on record where unnatural mothers have taken this means of disposing of the little victims of infanticide, but in these instances so much evidence of another character relating to the points at issue was produced, that conviction or acquittal was effected without dispute as to the disposal of the body, and hence no particular scientific notice was ever to any extent taken by those who had the matters in charge.

The theory of the defense in this case was, that the time in which it was alleged by the prosecution that the cremation of Mr. Druse's body was accomplished, was altogether too short to effect the result obtained — they, the defense, claiming that two or three days must have been necessary, while, according to the evidence offered by the prosecution, twelve to fifteen hours was the longest possible time the parties could have been engaged in the operation. The amount of fuel required, which they claimed would be enormous, was also brought prominently in question by the defense.

In anticipation of the efforts of counsel to question this part of the theory of the prosecution, and in the absence of sufficient recorded data upon which a satisfactory opinion could be predicated, it was deemed necessary to institute a series of experiments in the cremation of animal tissues. In this investigation Dr. Suiter was ably assisted by Dr. Theodore Deecke, special pathologist of the State Lunatic Asylum at Utica, who was associated with me as an expert witness at the trial.

In making these experiments especial care was taken to arrange the stove, as to its surroundings and exposures, to correspond as nearly as possible with the situation of the stove in the house where the alleged deed was committed — the same stove having been procured for the purpose, in which the body of William Druse was said to have been cremated.

The particular points to be determined upon were : the length of time that would be required to effect a complete incineration, and how much fuel would be consumed. Dry pine shingles and thoroughly dried sticks of pine wood were used as fuel in imitation of the alleged cremation, and both human and animal bodies were burned. The amount of wood used was carefully weighed for each experiment and compared to the weight of animal tissues cremated. A large number of experiments were performed in the manner indicated, and I regret that the limits of this chapter will not permit me to recount them in detail. In brief, as results of these experiments, we were able to positively state that a human body weighing 140 pounds could be reduced to a state of complete incineration in the remarkably short time of eight hours ; and that only one and one-quarter pounds of such fuel as we made use of would be required to each pound of mixed animal tissues.

It may not be amiss, before closing, to remark that the microscope was a very valuable assistant in the various stages of the investigation.

Stains found on the under surface of the floor-boards in the room where the tragedy took place, on the under surface of the hearth of the stove near the place where the head was severed from the body, and upon the walls in the room where the body was cut in pieces, were all proven by microscopical examination to have been produced by blood, the corpuscles of which corresponded as to form and measurement with those of human blood. The corpuscles from the different places mentioned appeared to be exactly alike and were undoubtedly from the same blood, which was in direct corroboration of the confessional testimony of the boy Gates. A gray hair found underneath the leg of the stove, which was situated near the place where the head was separated from the body, the microscope proved to be a human hair, and by its measurement it appeared to be a whisker. This, also, was corroborative of the Gates' testimony. The discovery of a few gray hairs and several small fragments of bone, in a sugar hut, led to the arrest and imprisonment of a man who was suspected of having been concerned in the disposition of the body. The materials were believed to have belonged to the body of Mr. Druse. Microscopic examination showed the hair to be that of the *Mephitis Americana*, or common skunk, and the fragments of bone found were from some of the lower animals. Upon reporting this, the grand jury failed to find a bill of indictment against the supposed accomplice, and he was released from custody.

A microscopic examination of the ashes proved that the wood, before it was charred, belonged to the conifera order of trees, which was corroborative of the confessional testimony given, to the effect that dry pine shingles were used in the cremation of the body.

CHAPTER XIX.

RESPECTING IMBIBITION OF POISONS.

PROF. JOHN J. REESE, M. D., says: In the remarks accompanying a paper by Dr. Geo. B. Miller on the *Post-mortem Imbibition of Poisons*, the very important and significant question is raised, or rather is quoted as having been "submitted to the chemists who were consulted" in a recent somewhat remarkable case of suspected arsenical poisoning: "Can you discriminate by chemical analysis, on an exhumation of the cadaver, between the arsenic we are sure to find therein, viz., that which was produced by absorption, *post-mortem*, in the embalming process, from the arsenic or other poison given before death, and which undoubtedly caused death?" Or, as it is afterwards put: "Can the chemist, by any means now known to science, detect the murderer who poisoned his victim, and at once thereafter fills the body with an arsenical solution, under the pretense of embalming it, using an extra large quantity purposely?"

The case of suspected poisoning which has given rise to the above queries is both interesting and instructive, and the circumstances attending it are strongly suggestive of guilt on the part of the accused. I am further of the opinion that this is by no means a solitary instance of such a criminal procedure. I will, therefore, with your permission, endeavor to reply to the above questions, seeking to throw such light as I may be able upon a confessedly obscure subject.

The subject of the *post-mortem imbibition of poisons* has not, I think, received from toxicologists and medical jurists that degree of consideration that its importance demands. Indeed, until within a very few years past it appears to have been entirely ignored. The distinguished Orfila, it is true, as far back as the year 1847, alludes to the fact of such imbibition, based upon his own experiments on animals, and he admits the possibility of its being practiced on the human body with a criminal intent; and Sir R. Christison also records his belief in the same possibility, although he had never himself heard of an actual case.

In the year 1877 I published a paper (Trans. Col. of Phys. of Philadelphia, 1877) on the *Post-mortem Imbibition of Poisons*, based on a series of experiments, made at my suggestion and under my supervision, by Dr. George McCracken, of Philadelphia, at the University of Pennsylvania, with a view to determine this question more specifically. These experiments, carefully conducted upon the bodies of dogs and cats with solutions of arsenious acid, corrosive sublimate and tartar emetic injected into their stomachs, conclusively established the fact of the imbibition, or osmosis, of these different poisons into the various organs, as the diaphragm, heart, lungs, liver, spleen, kidney and urinary bladder — these various substances being detected in the above-mentioned organs by the appropriate chemical reagents.

I do not think, however, that either the brain or spinal cord was examined at that time. The subsequent researches made by Dr. Miller, as detailed in his paper above alluded to, entirely confirm this statement; with the additional fact that traces of arsenic were discovered by him in the brain; and which later assertion seems to be corroborated by some experiments made by Dr. V. C. Vaughn, of Ann Arbor, Michigan, in 1883, and by Dr. Kedzie, of the Michigan Agricultural College, about the same time.

As regards the question whether a poisonous solution

injected into the human stomach after death, can penetrate by osmosis into the *brain* and *spinal cord*, although
the above experiments would seem to justify our assent
to the proposition, nevertheless the proof is not quite as
clear and positive as we could wish; since, in some of
these experiments, it is intimated that the solution, at
the time of the injection into the stomach through the
pharynx *regurgitated through the nostrils*, and consequently might possibly have gained access to the brain
directly, through the ethmoid and sphenoid cells. At
all events, I think that this later question, viz., the *post-
mortem* imbibition from the stomach into the brain and
spinal cord, should be more thoroughly investigated
before challenging our unqualified assent. I deem it to
be a question of such vital importance in this particular
connection, that its settlement should be established by
the most indisputable proofs.

The importance of this question will be perceived at
once. If, for example, it should be positively deter-
mided that the poisonous solution does *not* penetrate
into these two great nerve centres after death, we should
then be in possession of a certain and reliable means of
discriminating between a poison administered before
death, and one injected into the body after death; since
its discovery in the brain or spinal marrow of the deceased, by the toxicologist, would then be proof positive
of *ante-mortem* poisining; although, on the other hand,
its non-discovery in these organs, would not, of itself,
justify an opposite conclusion, viz., that of the *post-
mortem* imbibition, inasmuch as we well know many
fatal cases of poisoning do occur without the analyst
being able always to detect the lethal agent in these
organs after death. I would therefore again beg to
draw attention to the importance of further investigations of this vexed question, special care being taken to
guard against regurgitation into the posterior nares,
when the solution is injected into the stomach through
pharynx; and also recommending the additional experi-

There is an error in numbering these pages—384–387.
No matter is omitted

ment to be made of injecting the poison through the rectum into the intestines, since the imbibition into the different viscera will take place from these latter organs equally well as from the stomach.

In conclusion, I will endeavor to answer the final important and practical question in your article, in which the appeal is made " to chemists throughout the world : " "How can we discriminate, so as to detect and differentiate the poison taken *ante-mortem*, which caused death, from the poison purposely introduced after death into the abdominal cavity, or per rectum, for the purpose of hiding the crime under pretense of embalming the body?" I believe there is no known method by which such discrimination can be made, if we have to rely exclusively on the chemical analysis of the body, since this simply establishes the *fact* of the presence or absence of the poison, but does not necessarily disclose its *mode of introduction*. But the following points or propositions may, I think, materially aid in the elucidation of the subject :

I. A knowledge of the *symptoms* before death, where this is obtainable, will frequently throw much light on the case, although too much stress should not be placed upon symptoms merely, inasmuch as the symptoms of many diseases strongly resemble those of certain poisons. Thus, I have known several cases of fatal arsenical poisoning to have been mistaken for cholera morbus, and treated as such by the attending physicians, and the certificate of death to be so made out, but in which I subsequently detected in the viscera large amounts of arsenic.

II. The chemical examination of the urine of the deceased. As is well known, the kidneys rapidly eliminate arsenic (and the same is true of other mineral poisons) from the body ; and it is generally possible, for the analyst to detect the poison in the urine, both before and after death. Its discovery in this secretion may, I think, be regarded as very conclusive evidence of *ante-*

25

mortem poisoning; for, although it is true that the urinary bladder, in common with the other abdominal viscera, was found contaminated by the *post-mortem* imbibition of the poisonous solutions, in the experiments above detailed, yet I think it scarcely possible that the poison would percolate through the coats of the bladder so as to effect the contained urine. The record of the experiments was simply the production of a yellow spot on the surface of this organ. Nevertheless, in a capital case where the evidence of poisoning hinged upon this particular aspect of the subject, I should not like to swear that such a thing might not be possible.

III. The finding of the poison, on the exterior of the organs, and not in their interior, is, I think, very positive evidence of *post-mortem* imbibition, since in a true *ante-mortem* case the absorbed poison is always deposited in the interior of the organ quite as distinctly as on the outer surface.* But, practically, this is often a difficult matter to decide, especially after the lapse of a long interval of time, and where the organs have become much broken down by decomposition.

IV. The discovery of poison in the stomach after death cannot be regarded as absolute proof of its *ante-mortem* administration, since, as we have seen, it might have been injected after death; and there is also the further *possibility* that if introduced into the intestines through the rectum, it might by imbibition penetrate into the interior of that organ, just as in the supposed case of the urinary bladder above alluded to.

Philadelphia, July, 1887.

---

* *Vide* Orfila. *Toxicologie*, 1852. *I*, p. 63.

CHAPTER XX.

RESPECTING CIRCUMSTANTIAL EVIDENCE IN POISONING CASES.

JOHN H. WIGMORE, Esq., of the Boston Bar, says:

"*Homicidium quod nullo presente, nullo sciente, nullo audiente, nullo vidente, clam perpetratur.*"

The purpose of this writing is to examine a single one of the important topics of medico-legal investigation from the standpoint of the lawyer. Whatever has been said in this connection in the treatises on medical jurisprudence has been said chiefly for the benefit of the medical witness. Valuable hints to the advocate have been given by Wharton, Greenleaf and others, and Bentham and Wills have paid special attention to the peculiar features of circumstantial evidence in general. But in the realm of poisoning cases, at least, the subject calls for a more direct and specialized treatment than has been yet given to it. What follows will, it is hoped, at once explain and justify this statement.

It may be here pointed out, however, that poisoning cases offer a complication of questions and a wide range of investigation not present on the ordinary occasions when medical testimony is needed, and more capable of useful analysis. They differ, moreover, from the more common instances of homicide, for, on the one hand, the pleas of self-defense and of provocation, and other exculpatory issues are, in the nature of the case, almost impossible, and the issue is reduced to a question of murder or utter innocence, and, on the other hand, the evidence cannot range over a vast field of facts limited only by the possibilities of methods of destruction, but is limited to a certain class of agents. For almost every step of proof, moreover, medical evidence is needed.

In brief, there is a sameness about the evidence which admits of an induction and invites a helpful analysis and classification.

PROOF OF THE POISONING.

What, then, are the propositions to be proved in a trial for murder for poisoning, and what is the nature of the evidence which may come into play relative to each proposition? The question presented is a single one: Did the defendant kill the deceased by poison?" Let us separate this into its component parts, noting the evidence relevant to each part. We can then examine the validity and significance of the analysis.

In proving a charge that the accused killed the deceased by poison, these three propositions are involved:

First. That the deceased died by poison.

Second. That the poison was administered by the accused or by his agency.

Third. That the accused foresaw the harmful effects of the substance given.

1. The first proposition covers that part of the subject of proof commonly called *corpus delicti.*

This proposition is the first and all-important step, for without its establishment the case must entirely fail. Unless the deceased died by poison, it is unnecessary to inquire who poisoned him. It is this part of the case which, in poisoning trials, is raised into an importance far greater than in ordinary trials for murder. Usually, upon proof of an unnatural death, no further evidence as to the *corpus delicti* is needed, and no special difficulty attaches to the nature of the agency causing the death. But in this class of cases specific proof of the use of poison always remains. It is often the most serious task for the prosecution, and has, on many occasions, become the turning point of the case. Moreover, there are open only two modes of proving

THE CAUSE OF DEATH

to have been poison — proof by the results of analysis of portions of the body, or of substances a part of which has been known to enter the body, and proof from the observed symptoms and appearances, both before and after death. Not infrequently one of these modes may be inconclusive or may become impossible, and in that case the line of evidence is still more restricted.

CIRCUMSTANTIAL EVIDENCE.

It should be observed, too, that these methods both rest on circumstantial evidence solely. Direct evidence, it may be premised, properly includes all testimony immediately asserting or denying the existence of the fact to be proved, and thus, in order to be conclusive, calls only for an inference as to the credibility of the witness. Circumstantial evidence, on the contrary, is evidence tending to prove some other fact, presumably relevant (called by Bentham "sign," "*factum probans*," "evidentiary fact"), and thus, in order to be valuable, needs an intermediate inference or inferences other than as to the credibility of the witness. Testimony, then, to the symptoms, or to the reaction produced by certain portions of the body subjected to chemical analysis, is circumstantial evidence, but it needs the help of an inference to bridge the gap between these facts and the main fact of death by poison.

2. Passing to the second principal fact, it may of course be proved by direct evidence of the administration of poison by the accused. But this evidence can rarely be secured.

Practically, circumstantial evidence alone is available. We are relegated to proof of subsidiary facts, from the existence of which we may infer the existence of the principal fact. These subsidiary facts, or groups of facts, forming a complete chain of evidence, are four in number:

(a) Previous possession of the poisonous substance ;

(b) Opportunity of administration ;

(c) Antecedent possibility or probability (including motive, expressed intention, etc.); and

(d) Impossibility or improbability of administration by other agencies.

THE ADMINISTRATION OF THE POISON.

In the absence of direct evidence of administration by the accused the evidence offered must be calculated to prove these four facts. Exactly what probative force should be allowed to each or to the co-existence of all will be discussed.

3. Thirdly, it must be shown that the accused administered the poison with knowledge of its probable effects.

This issue does not frequently become important, for the evidence that goes to prove the second main fact will usually serve to prove this one. In any case the evidence available must, of course, be mainly circumstantial. It is only rare, however, that it is possible for the defense to make even a show of contending upon this point. In Miss Blandy's trial (at Oxford, 1792) this issue became vital, for the innocence of the accused depended upon the truth of her story, that she had administered the fatal powders to her father as a love potion to retain his affection. Again, in the case of George Ball (in 1860, at Lewes), it appeared that an overdose of prussic acid was given for medicinal purposes, and upon the fact that the overdose was an innocent mistake turned the fate of the accused.

GENERAL ADMISSIONS.

These classes comprise the largest share of the evidence that can be relevant to a poisoning trial. But there remains another and more important class of evidence bearing on the general question of guilt. This class comprises all general admissions of guilt; not admissions of one or more of the subsidiary facts above mentioned, for these take their proper place under the preceding classes of evidence, but all evidence in the nature of conduct or words subsequent to the guilty act and tending to show a consciousness of guilt on the part of the accused.

This evidence includes: First, express oral or written admissions of guilt, or what are called in criminal law confessions; and, secondly, conduct pointing toward a consciousness of guilt.

Under one or another of the above divisions it is believed that all possible evidence must fall. Let us now examine this analysis in detail, with a view of testing its validity and illustrating its significance.

DEATH OF THE DECEASED BY POISON.

1. In the first place, remembering that unless a death by poison can be proved, the case of the prosecution falls to the ground, care will be taken at every step to secure the complete and thorough establishment of this point. Remembering also that there are but two kinds of evidence available for this purpose, analytical and pathological, it will be ascertained, as early as possible, whether (from other circumstances of the case or from the nature of the poison used) either of these kinds will be in any degree unavailable. Some of the most important poisoning trials on record have turned entirely on the question whether the deceased died by poison. The Palmer case (Rugely, 1885-6) and the Lamson case (Wimbledon, 1882) are instances in which there was an utter failure of one sort of evidence, and the necessity arose of relying upon a single set of phenomena for proof.

ACONITIA.

In the Lamson case the symptoms spoken of by the deceased boy — a severe heartburn — suggested aconitia as the cause. Now, in the present state of our knowledge, the chemical tests for aconitia are unreliable and practically useless. Moreover, the symptoms exhibited during illness are far from conclusive. How, then, can aconitia be detected? The only trustworthy effects are two: its taste and its effect on small animals. Any substance containing aconitia produces a tingling and numbness when applied to the tongue or the lips, and when injected into the back of a mouse causes a characteristic staggering, paralysis and asphyxia. When these two tests

agree, the presence of aconitia is absolutely certain.
These tests accordingly were made with portions of the
fluids in the body of the deceased, and aconitia was
found in considerable quantities. A conviction ultimately ensued.

STRYCHNINE.

A similar instance is furnished by the Palmer case.
Strychnia was suspected as the cause of death. It is not
true, at the present day, that there is no infallible test
for strychnia, for if by the "color test," so called, a particular succession of colors is produced, the presence of
strychnia is determined beyond a doubt. Yet strychnia
in a fatal quantity may be so minutely distributed
throughout the system that the failure of this test to
produce the proper colors does not prove that the poison
is not present, and in that case resort must be had to the
symptoms alone. This was what occurred in Palmer's
trial. Whether through carelessness or through imperfection of the methods used, the body furnished no
certain evidence, chemically, of the use of strychnine.
The contest took place upon the significance of the symptoms, and the leaders of the medical profession were
marshalled on either side. In this case, as in Lamson's,
the cause of death was the crucial issue of the trial; for
if in truth the death was by poison, it was impossible
upon the rest of the evidence to suppose any hand
but Palmer's had administered it. Tetanus, epilepsy,
angina pectoris, were all suggested, but the symptoms
of strychnia were too clear and a verdict of guilty was
rendered.

POISON THE CAUSE OF DEATH.

These illustrations will suffice to suggest the very
evident moral. Let it be your foremost care, upon
undertaking a poisoning case, from the first step of
preparation to the final address to the jury, to establish thoroughly the fact of poison's agency in the
death. It may be said that this cannot be seriously
questioned by the defense. But the possibility of

a slip should always be guarded against most carefully.   If, as in the case of premature burial (Donnellan's case), or of a suspicion that a substance like
aconitia or strychnia has been the cause of death, there
is a possibility that evidence of one kind or the other
will fail, all efforts must be concentrated on the available.
Experts must be instructed and examined with a view to
leaving no opportunity for failure in this essential step.
It is needless to add that the defense, in the same way,
if there appears any possibility of the weakness in the
prosecution's case on this point, must seize upon it and
make good use of such a stronghold.

ADMINISTRATION OF POISON BY THE ACCUSED.

2. Passing to the second issue, let me take up and
illustrate each of the subsidiary facts separately, and
then say a few words concerning their probative force
considered jointly.

PREVIOUS POSSESSION.

(a) The evidence presented may constitute a strong or
weak case, according as it fixes possession upon the
accused, with details of time, place and material, corresponding more or less closely to the known circumstances
of death, and having a significance more or less exclusive
of innocent explanation.   For example, the evidence
may show as in Dr. Smethurst's case (Richmond, 1859),
no more than that the accused being a physician, and
constantly having in his possession drugs of all sorts,
might without improbability have possessed a quantity
of the fatal drug; or the evidence may be, as in Palmer's
case, of the purchase of packages of strychnia within
two days before the death; or as in Dr. Pritchard's case
(Glasgow, 1856), the prosecution may even be able to
show that some one poisoned tapioca, taken by the
deceased, tallies with a quantity of the same substance
found in the accused's room and tainted with the same
poison.   In Madeline Smith's case (Glasgow, 1857), she
asserted that the arsenic which she was proved to have
bought was used by her for the complexion, and this

explanation helped to diminish somewhat the effect of the prosecution's evidence.

NEGATIVE EVIDENCE.

It should be noted that the effect of much evidence will be simply to contradict the impossibility of previous possession. Such evidence does not purport to be inconsistent with lack of possession, and may, in fact, be perfectly consistent therewith. It is offered with the humbler, yet highly important, object of annulling possible efforts of the defense to show that possession by the accused was impossible. For instance, in Donnellan's case the prosecution, on the theory that laurel water, obtained by distillation, was the fatal agent, offered evidence that Donnellan possessed a still, by means of which laurel water could have been made. This was of no direct value to prove the subsidiary fact of previous possession, but it was of value to defeat any possible efforts of the defense to show that Donnellan did not have it in his power to obtain laurel water. The contrast between such negative evidence and evidence of a direct and affirmative character is illustrated by another circumstance of the same case. In Donnellan's library was found a single volume of the transactions of the Philosophical Society. The volume was uncut except at a single place, and at that place was an account of the method of distilling laurel water. Nothing could have been more important to indicate a preparation by Donnellan of laurel water.

It must be added that this kind of evidence, which may be called negative, is not peculiar to the subsidiary fact—previous possession—under consideration. It may be offered under any of the other heads under II, as, under (b), opportunity to disprove an alibi. It may also appear under (1) as, for example, the evidence used in Palmer's case to rebut the contention of the defense that the failure to detect strychnia in the body by chemical analysis conclusively proved that strychnia had not caused death.

OPPORTUNITY.

(b) Here also the evidence may have a wide range as to probative force. It may only show a general possibility of administration by the accused, or it may show specific and oft-recurring opportunities. In the Lamson case, it appeared that the accused, shortly before the fatal sickness, had called to see the boy and had given him a capsule, which the boy swallowed. Here the opportunity was proven with unusual detail. In the cases of Dr. Smith and Dr. Pritchard, the attendance of the accused during the last illness of their victims, and general supervision exercised by them over the medicines administered, placed beyond doubt the existence of continued opportunities. In every case where the accused is a physician or a surgeon (and these comprise a very large number of the recorded cases), proof of opportunity is not likely to be wanting.

Lack of opportunity is of course one of the strongest issues open to the defense. Success, for example, in proving an alibi ends the cause at once. But the existence of an alibi is not the only contention that can arise under this head. Attention must be called to a class of cases in which the requisite of opportunity is apparently satisfied, and yet closer examination shows the contrary. In the trial of Madeline Smith it was proved that the deceased must have swallowed 200 grains or more of the arsenic which caused his death; and it was pointed out that the successful administration, by one having a hostile intent, of so large a dose of arsenic was an extremely improbable and almost unheard-of occurrence. There was, therefore, strictly speaking, no opportunity, because no physical possibility, of administration by the accused, and one was forced to assume suicide. A similar instance is furnished in the trial of Adelaide Bartlett (London, 1886). Large quantities of chloroform had been taken by the deceased, not in the ordinary manner, but by swallowing, and great weight was given to testimony that the chances were enormously against the suc cessful administration of chloroform in that manner during sleep. The nature of the substance administered and the mode of administration may thus often afford valuable indications as to the feasibility of administration by the accused.

ANTECEDENT PROBABILITY OR POSSIBILITY.

(c) Under this head belongs that multifarious mass of evidence touching, on the one hand, the habits of the accused, his disposition and general character, his relations with the deceased, his business affairs and all other circumstances calculated to call into action a motive for or against the murder, and on the other hand, touching his previous intentions, expressed or implied, and his preparations, attempts, or threats, if any of these indications can be gathered from his words or conduct previous to the time of administration. In poisoning cases, however, the treatment of this class of evidence is not materially different from that demanded in ordinary cases of alleged homicide, and does not need further illustration. It is to be noticed that the fact of previous possession, the chief probative force of which is that it tends to identify the accused with the person administering the poison, may often have some additional force as indicating preparation on the part of the accused.

MOTIVE.

It may be added that motive (a convenient but abbreviated term for the circumstances calculated to call an emotion into play) is not in itself a neccessary thing to be proved. It is simply a most important one of the several sorts of evidence that go to make up a general antecedent likelihood of guilt. Neither the presence nor the absence of motive is conclusive. Here, however, as in the other subsidiary issues under II, the defense may theoretically prove an absolute negative ; that is, that the commission of the crime by the defendant would be entirely contrary to what would have been expected from his character and the circumstances of the case ; although, as in the case of evidence of lack of opportunity, this cannot practically amount to more than evidence showing a high degree of improbability.

THE IMPOSSIBILITY OR IMPROBABILITY

of administration by an agency other than the defendant's is always an item of extreme importance. It is of most value to the prosecution when strong evidence is introduced by the defense under other heads, leaving suicide, for example, as the only other tenable hypothesis. In such a case, if this avenue can be closed up by the prosecution, an amount of evidence on the remaining points, otherwise insufficient, may prove conclusive. For instance, in the trial of Madeline Smith, already spoken of, purchases of arsenic by the accused on three occasions shortly before the death were well proved, and a sufficient motive (the dread of exposure of a criminal relation on the point of her marriage) was not wanting. It appeared further that the deceased came to Edinburgh, the scene of the fatal occurrence, at nine o'clock in the evening, intending to meet the accused, and was not seen or heard of again until two o'clock the next morning, when he returned to his lodgings with the fatal illness upon him. The remaining evidence was such that the only other hypothesis possible was that of suicide. Here the prosecution was unable to close the gap. The theory of suicide, indeed, was not an improbable one in itself, and thus the evidence tending to show lack of opportunity was enabled to have its full effect. So, too, in the Bartlett case, already mentioned, the evidence as to the physical impossibility of forcing chloroform, in the quantity allged, down the throat of a sleeping person, practically threw upon the prosecution the burden of showing that suicide was impossible, and as in fact this was not at all unlikely, the defense stood in a very strong position.

GENERAL ANALYSIS OF THE SUBJECT.

These illustrations, with the comments thereon, have been premised for the purpose of paving the way for the reception of two important statements concerning the analysis which has been presented. In previous discussions upon the nature and use of circumstantial evidence, including those of Bentham and Wills, the treatment of the subject seems to have been confined to a general enumeration of the different lines of evidence and the particular probative tendency of each. We do not find satisfactory attention given to the possibility of classifying the component subsidiary facts of circumstantial evidence accurately and exhaustively, or to the relative importance of the different subsidiary facts or groups of facts, and, in consequence, the profound discrimination and subtlety exercised in these dissertions fails to afford us the maximum of usefulness. Whether, in the subject of circumstantial evidence at large, such coherence and logical relations exist, need not be here considered ; but, confining ourselves to circumstantial evidence of the administration of poison, it is believed that this evidence is characterized by such qualities, and it is hoped that they are clearly and correctly brought out by the analysis which has been offered. Let us see, briefly, whether in the evidentiary facts above discussed (a, b, c and d, under II), there does not appear an exhaustiveness and a certain logical coherence.

1. By exhaustiveness (for want of a better term) is meant that under one or another of these heads must all circumstantial evidence come which tends to prove the act of delinquency on the part of the accused.

Leaving out admissional evidence and direct evidence of administration (which is, of course, rare), we have the certainty, in preparing and presenting our case, that every item of evidence secured or sought for has for its object the proof of one or another of these facts (a, b, c and d above), and gains therefrom its entire significance and value. It is hardly possible in this place to demonstrate that this analysis is in fact complete and exhaustive, partly for the reason that it is impossible to prove a universal negative. Let the reader test the analysis for himself. It rests upon an examination of the process of

ratiocination, natural and necessary in proving the fact
of administration, and upon the fact that no instance
has been discovered by the writer which does not fall
into one or another of the above classes. If there is
evidence amply and explicitly establishing each of these
separate points, the main fact follows as an unavoidable
culmination. Such complete array of evidence, however,
rarely appears, and the question arises, what probative
force, in the absence of complete proof upon all the
points, is to be given to such evidence as exists? This,
of course, brings us to the question, What is the relative
probative value of each subsidiary fact, that is, to a consideration of the logical coherence of these facts?

PROBABILITIES.

2. It is, of course, impossible in such a matter to look
for mathematical relations, or to expect to find that veri-
fication of some formula of variables. The tendency to
construct an artificial theory upon facts which do not
admit of it must carefully be restrained. But certain
uniform characteristics may be pointed out. In the first
place, complete proof by the defense of the absence of
any of these links is fatal to the case of the prosecution ;
that is to say, it is necessary for the prosecution that
each one of these facts should at least not be impossible.
The impossibility of Capt. Donellan's making or pro-
curing laurel water, if it could have been proved, would
have ended the case. Such absolute proof can of course
rarely be furnished, and the defense must be content
with making every effort to secure it. The illustrated
cases already cited will make this proposition sufficiently
clear, and render unnecessary further examination in
detail. Suppose, however, that upon three of the points
the prosecution brings forward complete evidence, while
on the fourth point the evidence is equally balanced. In
such a case the mind seems to find no difficulty in
reaching a conclusion upon the remaining evidence and
in inferring the existence of the main fact.

Again, if there is satisfactory proof upon two of the

points and upon the other two the evidence is equally balanced, the existence of the main fact can never be considered as proved. The writer does not recollect any cases, however, which illustrate the last two propositions. To travel further through the different possible combinations and to examine the significance of each, would not in this place be expedient. Enough has been said, it is hoped, to show that there is between these subsidiary facts, or groups of facts, a certain coherence and complemental relation. It is impossible, as has been said, in matters of inference to arrive at a mechanical certainty, or to measure to a fraction the weight of ingredient arguments. Perhaps in marking out on the one hand, that amount of evidence which certainly will not sustain an inference of guilt, and, on the other, that amount of evidence which certainy will not sustain such an inference, we have gone as far in quantitive analysis as the nature of the materials will permit us safely to go. My object is gained if I have shown that these linked facts make up a logical whole, the component parts being mutually so related that a variance in one materially affects all the others.

THE VALUE OF THE ANALYSIS.

We have now concluded our examination of this analysis; and the question arises, of what value is it?

I omit any consideration of the scientific value which attaches to every truth, great or small. I omit, also, any argument as to its value in a purely juristic sense — as to the light, for example, which it throws on the vexed question how far evidence should be admitted of previous administration of poison to other persons by the accused. I refer only to its practical value to the lawyer in the administration of justice; and this is twofold:

1. Any sound analysis and classification must be of value in enabling the advocate to prepare his case intelligently. The search after evidence, the comprehension of its worth when found, the understanding of its proper

place in the order of proof, and of the relative force of the different pieces of evidence — all these must depend largely upon the correctness of the plan of campaign which has been formed by the advocate. There will be a certainty and a confidence — an ability to gauge each fact in evidence and to make the best use of it, such as could not exist for one working without some such map before him. Its usefulness in this respect, it is believed, will suggest itself without further illustration.

2. But chiefly the advantage to be gained lies in the fact that what has been here explicitly stated is nothing more than what is realized and acted upon by every juryman without a distinct perception of the underlying reasons. I mean that the analysis which has been offered is valid, because it is based on the actual logical processes followed by the mind. and expresses in terms that upon which the judgment of the juryman is unconsciously founded. Most of us could detect a logical fallacy without hesitation, where to explain the reason of it would perhaps be impossible. So a juryman seizes the weak and strong points of a case without understanding the logical basis of his judgment. The important result is that the advocate, forearmed with an analysis of the evidentiary needs of the occasion, anticipating, on the one hand, the difficulties which the twelve will instinctively feel, may the more effectively endeavor to remove or alleviate them, and forecasting. on the other hand, the points most likely to tell favorably upon the minds of the jury, may direct his energies towards emphasizing and enforcing them. It would seem that without some such analysis and classification as has been the subject of this writing, the efforts of the advocate must be in the dark, formless and incapable of producing their normal and best effect, whether in the preparation of a poisoning case or in its presentation to the jury.

26

CHAPTER XXI.

The Psychological Aspect of the Case of Prendergast, who Assassinated Mayor Harrison, of
Chicago.

The case of Prendergast, and all others of his class
who come under the head of "Cranks," is very important to society on account of the special criminal neurosis
which they possess. Prendergast is an instinctive criminal, like the rest of his class, because the gratification
of his urgent and absorbing self-regarding instincts is
not, and cannot be, controlled by any sense of duty to
others, of which he is congenitally destitute and consti-
tutionally incapable. The great question in his case is :
Has he lost the power of control over his thoughts, feelings and acts? If not, he should suffer the extreme
penalty of the law, as he is a sane man. If his senses
bear truthful evidence ; if his understanding is capable
of receiving that evidence ; if his reason can draw proper
conclusions from the evidence thus received ; if his will
can guide the thought thus obtained ; if his moral sense
can tell the right and wrong of any act growing out of
that thought ; and if his acts can at his own pleasure be
in conformity with the action of all these qualities, then
he is a sane man and should suffer the extreme penalty
of the law for his brutal murder of Mayor Harrison. If
Prendergast knew the act he was committing to be
unlawful and morally wrong, and had reason enough to
apply such knowledge and be controlled by it, any defense
of insanity should not save him from just punishment.
  In every given case, although the accused may be

laboring under a defective mental organization, if he still understands the nature and character of his acts and its consequences, and has a knowledge that it is wrong and criminal, and a mental power sufficient to apply that knowledge, and to know that if he does the act he will do wrong and receive punishment, and possesses a will sufficient to restrain the impulse that may arise from a diseased mind, such defective mental organization is not sufficient to exempt him from responsibility to the law from the crime. We find *occasional or accidental criminals* who do not present in form, feature or cerebral structure anything whatever to distinguish them from persons who have not been convicted of crime. They do not differ from others either in the quality of their intellectual or moral structure. They differ from them only in that they have given way to temptation in the circumstances of their life and calling. In these cases of course we should not expect to discover in the criminal brain any structural peculiarities. We have the *essential or instinctive criminal* who is so by reason of defective mental organization, the defects being of the nature of intellectual and moral weakness, which, being irremediable, lands its victim in the class of habitual criminals; of this sort are the vagrants, who, unable to settle to steady work or unfit to do it, are addicted to petty thefts, to acts of wanton mischief, to arson and to sexual offenses, natural or unnatural ; some of them are capable of even committing murder, whether it be in the gratification of a brutal passion, or in imitation of a murder which they have heard or read of, or in a stupidly indifferent way without any other reason than a sullen, gloomy and disquieting feeling in themselves, impelling them to its discharge in an act of violence. They are persons who, having the organic appetites and animal passions of human nature, lack the developement of reason and moral feeling by which these are surmounted and held in check in rightly constituted persons ; and the

consequence is that in face of the complex conditions of
civilized society, they are only able to fulfill their natural
instincts by acts that are necessarily anti-social. The
loss of the sense of moral relations, and the fine qualities
of reason that are corollate with it, does not necessarily
preclude intellectual acuteness of a low order, for the
weak powers of attention and reflection which these persons
possess, being engaged entirely in the narrow sphere of
their immediate interests, are sometimes shown in a sharp
cunning. Structural defects of brain are not always to be
found in these cases. There may be an insufficient com-
plexity of convolutions or other defects, but generally the
structures involved in the mental deterioration are too fine
and complex to enable us to detect and fix the physical
defects or derangements of them. Lack of intellect and
lack of moral feeling do not necessarily go together in
defective mental organization. In some instances the
defect seems to be mainly moral, the intellect not being
notably affected. It is remarkable how complete the
absence of moral sense may be while the other powers of
mind are little, if at all, below the average. Those who
labor under this moral deprivation are generally either
born of criminal parents, open or secret, and bred in a
criminal atmosphere, or they are sprung from families in
which insanity, epilepsy or some nearly allied neurosis
has existed, when they still own an anti-social inheritance,
although in this case due to disease. It is these two
classes only, comprising persons born with little or no
moral sense, that have a special criminal neurosis and are
instinctive criminals.

There is another class of criminals: *criminals due to*
positive disease. To this class belong the general paralytic, who commits thefts or other offences; the epileptic, who commits homicide while in the strange state of
abnormal consciousness that precedes or follows his
fits; the maniac and the melancholiac, who sacrifices the
life of others under the sway of overpowering delusions;
the case of moral or emotional insanity proper; psycho-

sensory insanity, as my friend Dr. Charles H. Hughes first appropriately named it, who, having undergone an entire change of character after an attack of insanity or of paralysis, show themselves as depraved as they were formerly virtuous. These and others of like morbid kinds constitute a distinct class in which the so-called crime is properly the accident of the disease.

Society must punish crime, not for revenge, but to protect society from the evil deeds of the criminal by penal restraint of him, and to deter others by the example and him by the memory of his sufferings, from repetition of his acts. Of course, the punishment of an insane person or an idiot is no benefit to him, as he cannot learn by it, while the example of its infliction, so far from serving as a warning to others, is an outrage to social feeling and brings discredit on the administration of justice. As I have said in the preface to this work, it is important that there should be made a close investigation and exact definition, first, of those crimes that are done by persons suffering from positive disease, such as insanity and epilepsy; and, secondly, of those forms of defective mental organization which are the results of a bad inheritance. The former I have endeavored to study clinically in this work, in the Medical Jurisprudence of Insanity, to some extent; the latter have not yet been seriously investigated, because the large and important material which exists in prisons has not been made scientific use of. What is needed now, is full and exact investigation of cases of the kind by systematic and painstaking observation of their hereditary antecedents, their mental and bodily characters, the conditions of their training and the exact circumstances of their crimes. Such biographical records would serve as very valuable instances for the formation of sound inductions, and thus lay the foundations of such positive knowledge as science might present with confidence for the instruction and use of those who make and administer the criminal laws. Our prisons should be used for the advancement of knowledge and for the building up of an individual psychology which would entirely displace the speculative theories that to-day, as a pseudo-psychology, is not of the least practical use to mankind, either in the breeding of children, the guidance of education, or the conduct of life.

INDEX.

A.

D

E

F

G

S   ,

www.ingramcontent.com/pod-product-compliance
Lightning Source LLC
Chambersburg PA
CBHW021343210326
41599CB00011B/733